Published by Passenger Transport Intelligence Services Limited

83 Latimer Road, Eastbourne, East Sussex, BN22 7EL
Telephone: 01729 840756
e-mail: info@passtrans.co.uk
Web: www.passtrans.co.uk

ISBN 978-1-898758-18-1

Author: **Chris Cheek BA FCILT MCIM**

This report is the latest in a series published as part of the *Bus Industry Monitor*
project originally launched by The TAS Partnership in 1991 and published since 1994
by this company (formerly called TAS Publications and Events Ltd). The reports are
researched, written and published by as part of the company's normal commercial
activities and are not funded, supported or sponsored in any way.

A *Bus Industry Monitor* Report

THE BUS DEMAND JIGSAW

Contents

Tables

Figures

Preface

Welcome to this latest report in the *Bus Industry Monitor* series. This volume focuses on demand for bus services and sets out to examine recent and longer-term trends in the context of the other demographic, economic and social changes that are taking place in society.

At the time of writing, the crisis induced by the COVID-19 pandemic has overtaken everything else in public policy terms and in the public consciousness. At this stage, we know that the society that will emerge from the crisis will be changed. We cannot yet know the nature and extent of those changes, nor of the economic impact that the virus will have in the medium to long term.

We can expect some changes to be to the detriment of the bus industry – accelerated home working and online shopping are two examples which would reduce passenger demand. This reinforces the point which I have long argued, namely that the forces governing the market for bus services are, to a remarkable extent, beyond the control of its managers. This is partly to do with the fact that people's use of bus services is *derived demand*: most people do not travel for the sake of it, as they might buy a book or music track; they only travel as a means to an end. To get to work, or the shops or to school or college, for instance.

When the reason for that journey is taken away, as for example when people bought televisions in the 1950s and stopped going to the cinema so often, the journey itself will not happen, and demand will fall. We can see that currently in the fall in journeys to visit High Street shops and other retail outlets.

If we can understand all those reasons for travel, and at the same time get to know what makes people choose one means of transport (or "mode") over another, then we are in a more powerful position to plan for the future and to do what we can to influence those decisions. This need is even more important as we seek to equip the industry for the new world which will emerge later in 2020.

That the decisions are capable of being influenced is surely beyond doubt. How else could one account for the remarkable difference in ridership levels between different parts of the country – even when areas are otherwise remarkably similar in economic and demographic terms.

Factors of leadership, overall quality, staff communication and motivation are all important in determining how successful local networks will be; the public policy context is, of course, vital, and can be seen in the success of different networks around the country.

In Chapter 2, we examine the context in which all this takes place: the classic marketing mix analysis is still a powerful tool in understanding all the influences, and in crafting a response.

Having done this basic work, it is possible quantify many of the influences on demand and thus estimate their importance to the overall picture. Looking at these in a series of disaggregated markets also enables us to see how different trends in different areas have influence diverging patterns of demand. Thus, the same model can predict growth in London or decline in other conurbations. It can also highlight areas where shifts have occurred which are larger than expected, and where further investigation and understanding would be merited. Examples include the remarkable downward shift in the bus market in the South West and Strathclyde region of Scotland in recent years; or the damaging loss of more than a quarter of paying passengers in Wales since 2010.

It would be easy to become deeply pessimistic about the future of the industry in the context of ongoing and rapid social change, and the seemingly unstoppable onward march of the private car. The Department for Transport's 2018 Road Traffic Forecast, which we discuss in Chapter 14, envisages further growth in the number of cars over the period until 2050, and further slowing of journey times as congestion grows, so driving bus costs up and demand down. The prospects for our High Streets do not look good in the face of the huge growth in online shopping, banking and other services, reducing demand for travel to shop by bus by almost 25% since 2010.

This pessimism must be set against the achievements of some sections of the industry over many years. This evidence suggests that companies can still survive and prosper in the future, provided that they understand the market context in which they operate, and act to deliver high quality services that are well-marketed. Communities can still enjoy high quality, frequent and sustainable bus networks – provided that their authorities provide the right conditions in which operator can deliver for their customers. That means good quality infrastructure, better-managed highway networks, the right parking policies and fair reimbursement for concessionary travel. It follows that the two working together – operators and authorities – can, and in many places do, deliver huge benefits for all concerned.

Thus, the analysis presented here is not purely deterministic; it is possible to influence stakeholders to introduce bus-friendly policies, and it is possible to improve service quality and to change public perceptions. It would also be a great mistake to set networks in aspic, adopting a framework in which change was slow, cumbersome and governed by political considerations.

Bus services do not exist in a continuum in which everything remains the same: change is something that operators always face – after all, something like 10% of us move house every year, roughly one third of the student population graduates every year and moves on, just as another third arrives to start their academic career. Children grow up and change schools; people change jobs, firms move and technology improves communications and empowers people in all sorts of ways. New houses are built, or old ones are converted and redeveloped. And, as we have already seen, shops close and are replaced by new ones (or sometimes not).

The hope of every government which seeks to reform the industry is to end the decline in bus patronage, and ensure future growth; on this count, every piece of legislation since the 1968 Transport Act has failed. There is little evidence to suppose that the fate of the 2017 Act will be any different. You cannot legislate to reverse decline or to change the public's behaviour.

In reality, then, legislation is largely irrelevant. What matters is understanding and influencing the travel decisions made by millions of individual people every day. Hopefully, this volume will contribute to that understanding, and to our ability to influence people's travel choices.

Chris Cheek
Passenger Transport Intelligence Services
April 2020

Figure 0-1: Boundaries for Area and Regional Analysis

HIGHLANDS, ISLANDS & SHETLAND

[Orkney and Shetland not shown]

Eilean Siar [Western Isles]

Highland

Grampian

NORTH EAST, TAYSIDE & CENTRAL

Tayside

Fife

Central

SOUTH EAST SCOTLAND

Lothian

Borders

Strathclyde

Northumberland

NORTH EAST

Tyne & Wear

Dumfries & Galloway

Durham

Teeside

SOUTH WEST & STRATHCLYDE

Cumbria

North Yorkshire

YORKSHIRE & THE HUMBER

NORTH WEST

East Yorkshire

Hull

Lancashire

West Yorkshire

North Lincs

NE Lincs

Merseyside

Greater Manchester

South Yorkshire

EAST MIDLANDS

Cheshire

Derbyshire

Lincolnshire

Notts

Staffs

Norfolk

EASTERN

WEST MIDLANDS

Shropshire

West Mids

Leicestershire

Cambs

Worcester

Warks

Northants

Suffolk

Hereford

Beds

WALES

Bucks

Herts

Essex

Gloucs

Oxon

Greater London

West of England CA

Berkshire

Wiltshire

Surrey

Kent

SOUTH WEST

Somerset

Hampshire

West Sussex

East Sussex

SOUTH EAST

Devon

Dorset

Isle of Wight

Cornwall

Chapter 1: Introduction

1.1 About this Report

This report analyses the market for bus services in the UK. It shows how demand has changed in recent years, identifies the reasons for those changes and considers how that might influence the future.

This introductory chapter sets the scene, providing an historic perspective on demand for bus services, looking at the long-term trends. A more detailed look at performance since 2005 then follows.

Chapter 2 then sets out the marketing mix for the bus, setting at the various influences on demand for bus services. Chapters 3 to 7 then look in more detail at the statistical evidence for those influences, examining price and revenue, demographics, concessionary travel, competition and journey purpose.

Chapters 8 to 12 examine the level of demand for bus services in different geographical markets, with detailed analysis of the markets the PTE areas, English Shires, Government Office Regions and the areas of devolved Government in Scotland and Wales and London.

Chapter 13 seeks to draw the strands together and considers demand outcomes in the light of all the changes recorded since 2010. Finally, Chapter 14 considers the prospects for the future, thinking about the expected future growth in car ownership and traffic levels and offers a more detailed look at the impact of the COVID-19 crisis.

1.2 Understanding Demand

The need to understand the components of demand and how they related to one another was first became apparent at The TAS Partnership in the early 1990s, in the form of a commission to examine the consequences for bus demand of the introduction of some form of road pricing. This was followed by the preparation of evidence for the House of Commons Transport Committee's 1994 inquiry into bus deregulation and its consequences.

Then came commissions for further work for the Department for Transport during the later 1990s and just after the turn of the century, ultimately leading to the development of the National Bus Model for the Department and subsequently for the Commission for Integrated Transport.

At an early date in the *Bus Industry Monitor* cycle, we started to include some of the conclusions and the statistical evidence from that work, and this has since evolved into this report.

It became clear that the influences on demand were many and various – a complex web, in fact, of inter-related issues, such as overall prosperity, the state of the employment market, who owned and was buying cars, changes in shopping habits and retail patterns, shifting trends in leisure activities, and very many more.

It was a bit like doing a jigsaw: identifying all the pieces and their shapes, trying to see how they all fitted together and finally assembling the whole picture. Twenty-five years on from that early work, jigsaw is still not was entirely complete – especially as various of the pieces themselves frequently evolve and change shape. However, the work done over the years does give a good understanding of how the overall picture looks. A version of the jigsaw as it stands today is therefore presented in Figure 1-1 below.

The various factors are grouped by colour – with the green pieces representing the various attributes of the bus product and the red pieces representing competition from other modes of transport (most notably the private car). The demographic influences on demand are coloured blue, whilst the state of the local and regional economy is coloured purple, whilst grey represents the various reasons people have to make their journeys (known as "journey purpose") and – increasingly – the factors which might encourage them to stay at home and not travel at all.

Figure 1-1: The Bus Demand Jigsaw

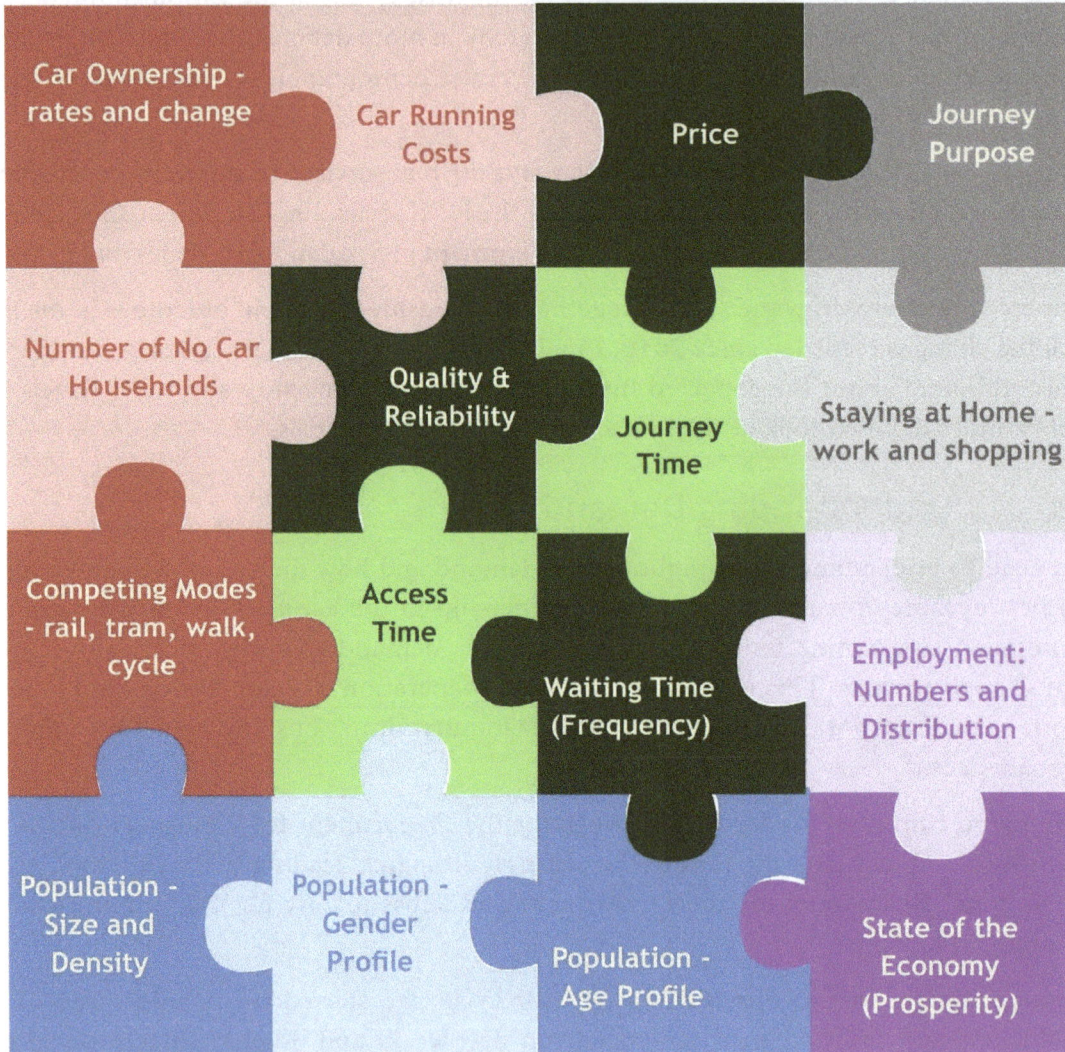

© PTI Services Limited 2017

Fundamentally, then, this report is about this jigsaw – taking the pieces apart and examining each one in turn, reviewing the statistical evidence for many of them and trying to show how they fit together again.

The first exercise is to put some perspective on the whole issue, by examining the history of bus demand since 1950 and how its market share has evolved over years since.

1.3 An Historic Perspective

1.3.1 The Industry's Decline

The market for bus services in the Great Britain saw a long period of decline between 1950 and the mid-1990s. In 1950, 16,445 million journeys were made by street-running public transport (in those days encompassing bus, tram and trolleybus services), according to statistics published by the DfT

By 1971, this figure had halved to 8,153 million. By 1993/94 it had halved again, to only 4,500 million. During the period from 1971 to 1998/99, only four years saw an increase of any size. Then in 1994/95, things began to change, particularly in London, where some modest growth got under way. Nationally, demand fell to 4.3 billion passenger journeys in 1998/99: this proved to be the industry's low point. From that position, and thanks to a variety of factors we shall discuss elsewhere in this volume, demand recovered to reach a peak of 5.27 billion in 2008/09, just as the major recession hit. Since then, decline has resumed, and even reached the previously buoyant London market after 2014.

The story of the period since 1950 is illustrated in the chart at Figure 1-2 below.

Figure 1-2: Bus Patronage since 1950

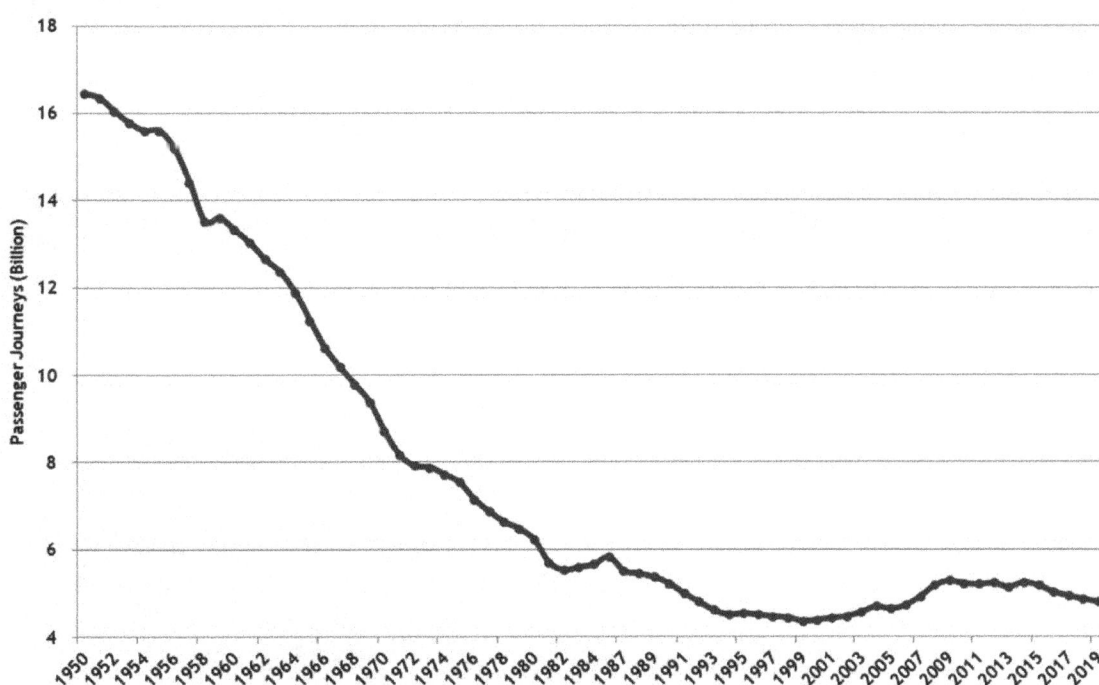

1.3.2 The Industry's Market Share

At the start of the 1950s, bus and coach services enjoyed a dominant position in the market for travel, with 42% of total demand in 1952. This compared with 18% for rail systems, 27% for cars, vans and taxis, 11% for pedal cycles and 3% for motorcycles.

By 1960, as the first motorways were opening and the mass market for car production was fully into its stride, bus and coach's market share had fallen to 28%. Passenger kilometres travelled by car had doubled in eight years and stood at 139 billion: this accounted for 49% of the total for that year, 282 billion.

In 1974, the rise in oil prices and the economic problems brought about by double-digit inflation prompted the first pause in the growth of demand for car travel since 1952. By

13

this time though, the number of passenger kilometres travelled by car had more than doubled again. It stood at 333 billion, accounting for 76% of the total, which had now reached 441 billion. Meanwhile, the share for buses and coaches had dwindled to only 14%.

By the time Iraqi invaded Kuwait in 1990, triggering another spike in oil prices, the bus industry's market share had halved again to 7%. Meanwhile, the private car had achieved a market share of 85%. At the same time, total demand had reached 690 billion. This meant that the total distance travelled by car had increased by a factor of ten since 1952 and stood at 588 billion kilometres.

After a pause at the end of the 1990s, car demand resumed its upward path, reaching a new record of 674 billion kilometres in 2007. Thereafter, the onset of the recession and other social changes drove the number back down. It fell back to a low point of 641 billion in 2013 before beginning to recover as the economy picked up once more, and petrol prices began to fall. The latest provisional figures for 2018 show that the distance travelled by car reached 672.7 billion.

Meanwhile, the market share enjoyed by buses and coaches was remarkably stable from the early 1990s onwards, maintaining a level of around 6%. However, in 2005, a revision of the methodology for estimating bus and coach demand then revised the figures downwards, and it stood at around five per cent for several years, before dropping again. The most recent figure was 4.4%.

A graph illustrating the historic trends is shown at Figure 1-3 below, whilst Figure 1-4 provides a snapshot of the market shares, comparing 1952 with 2018.

Figure 1-3: Demand for Travel by Mode 1952-2018

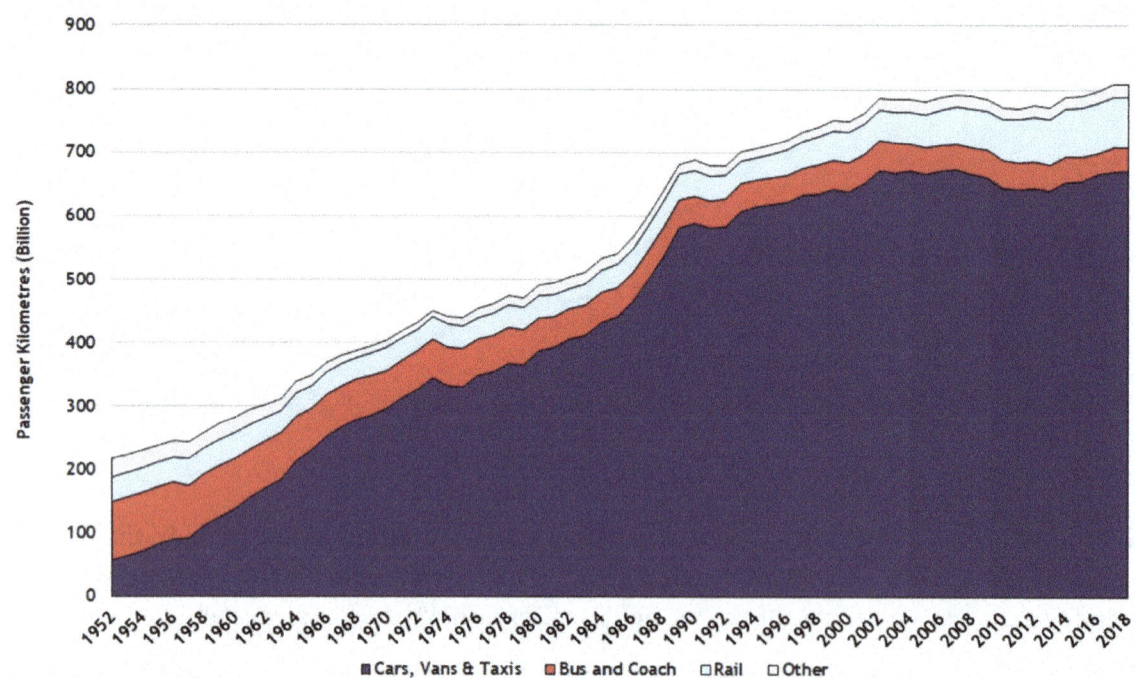

Source: Transport Statistics Great Britain, DfT

Figure 1-4: Snapshots of Modal Split, 1952 & 2018

1952

2018

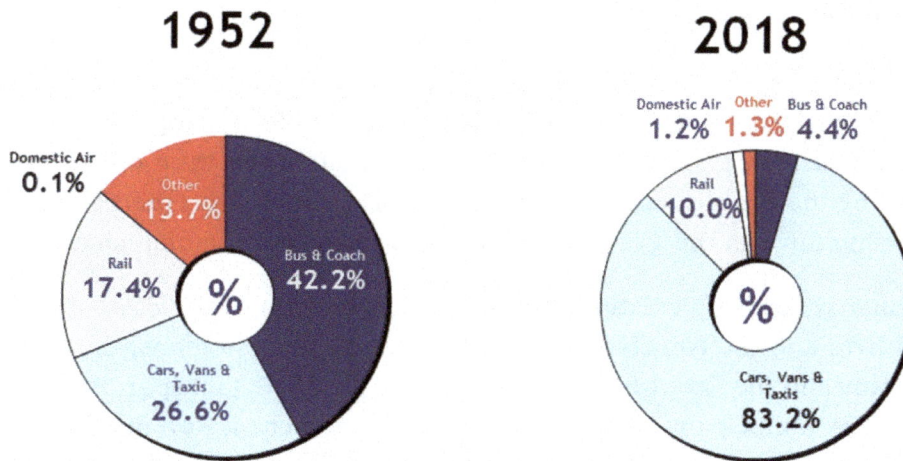

Source: PTIS analysis of Transport Statistics Great Britain, DfT

1.3.3   The Industry's Response to Decline

The first response during the 1950s was to increase fares, and they have gone on rising substantially in real terms ever since. In 1950, the average fare paid by passengers in today's prices was 24.7p. By 1960, this had risen to 29.3p. In the next fifteen years, the figure increased by 57% to reach 45.9p. By 1985/86, the year before deregulation, the average fare was 51.9p. Ten years later, this had risen to 67p. The most recent available figure is 102p.

Of course, average fare does not take account of changes in journey lengths and other market alterations, but it does provide an indication of the movement in pricing levels. The average fare paid is three times higher in real terms than in 1950 and 1.5 times higher than before deregulation.

The other 'standard response' was to cut services. Though there were some reductions in kilometres operated, this did not follow demand to the same extent. For example, between 1950 and 1965, demand for local bus services fell by 23.6%, whilst the number of kilometres run fell by 10.8%.

At the low point of supply in 1985/86, the number of vehicle kilometres stood at 2,077 million, which was 39% below 1950 levels. Service levels then began to climb again, reaching a post-deregulation high of 2,670 million in 1999/2000.

Although supply fell back again between 2000/01 and 2005/06, it began to increase again following the introduction of free concessionary fares to all English authority areas in 2006/07, until the onset of the recession, so that it reached a peak in 2008/09. Since then it has fallen back – largely driven by the reduction in supported services outside London. As a result, supply in 2018/19 was below levels seen in the early 1990s and early 2000s, though it remained 12.5% ahead of the pre-deregulation figure.

This means that although the number of people using local services has declined by 62.3% since 1960, the mileage operated has declined by only 26.5%.

1.3.4   Organisational Changes

Since 1930 there have been four basic forms of organisation and regulation within the industry, namely:

- The 1930 Road Traffic Act system, between 1950 and 1970, under which the industry was largely self-supporting, but protected from competition by the Road Service Licensing system. Until 1967, approximately one-third of the industry was owned by the private sector (mainly the BET Group), with the balance in public ownership through the Transport Holding Company or the municipal operators

- A similar system of regulation, but with the formation of the Passenger Transport Executives and the two state-owned specialist bus corporations, the National Bus Company and the Scottish Bus Group, under the 1968 Transport Act. Outside the PTE areas, municipally owned operators were untouched. Revenue support outside PTE areas was limited to some rural services at the margin. This system ran from 1970 to 1974

- The period of local authority involvement in services and revenue support ushered in by the Local Government Act 1972, which lasted from 1974 until 1986. By that year, 92% of local bus service kilometres were operated and 97% of passengers were carried by publicly owned enterprises. The total bill for industry support in 1984/85 reached £998m (£2,102m at current prices) and increased by a factor of almost five in cash terms between 1977/78 and 1984/85.

- The deregulation and progressive privatisation from 1985/86 to the present day.

Table 1 illustrates the decline in terms of the total number of local bus, trolleybus and traditional tram journeys lost since 1950 in each of these broad time periods. Figures exclude London.

The figures are also shown for the last five years to show how patronage has changed in the most recent period.

In terms of the total number of journeys lost each year, the deregulated regime has performed rather better than previous systems. This is shown by the average annual rate of decline, which was lower after 1986. There is a new break in the series following the DfT's revision of its patronage estimation methodology in 2004/05.

The introduction of a deregulated market did not reverse the decline in patronage, as the then Government hoped. However, neither did it lead to a significant worsening, and indeed the 20 years following the change saw the rate of decline virtually halved. Between 2005 and 2010, patronage rose thanks largely to the introduction of free concessionary travel in those parts of the country which did not already have it.

Since 2010, the combination of recession, cuts to funding and other social changes have seen demand falling once again, with the rate of decline now matching the 1950-1970 figure of two per cent per annum. A graph highlighting the phases and showing demand trends both outside and inside London is shown at Figure 1-5.

Table 1: Patronage Changes outside London under Regulatory Regimes

Time Period	Total Loss or (Gain)	% Lost/(Gained) during period	Average Annual Rate of Change
1950-1970	5,725	44.3%	-2.0%
1970-1974	941	13.1%	-3.1%
1974-1986	1,577	30.6%	-2.2%
1986-2005	1,397	32.2%	-1.3%
2005-2010	(145)	(5.1%)	1.0%
2010-2019	250	12.7%	-2.0%

Figure 1-5: Passenger Trends under Different Regulatory Regimes

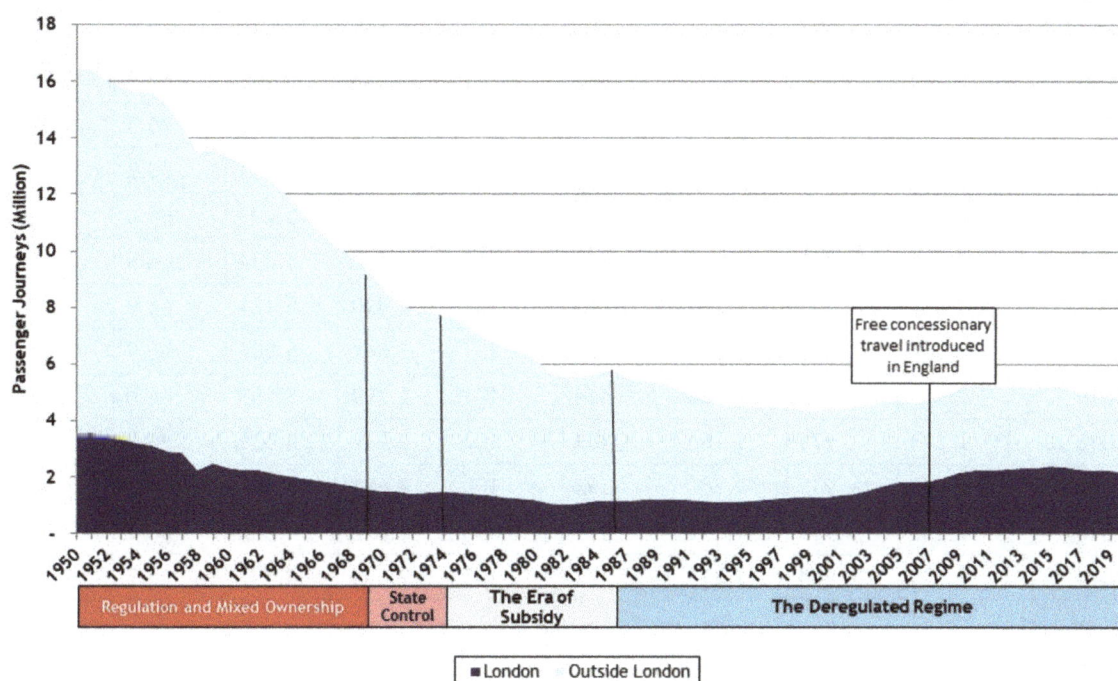

1.4 Recent Trends in Demand

As already noted, demand continued to fall after deregulation, until it reached a low point of 4.35 billion journeys in 1998/99. This represented a 21% fall since 1986/87 (29% outside London).

The period between then and 2004/05, the year of the interruption to the time series caused by a change of methodology at the DfT, it had recovered by 8.9% (though declining by a further 1.5% outside London). Thereafter, demand grew steadily, so that by the time of the onset of the recession in 2008/09, demand had grown by 13.8% nationally, and 7.5% outside London. The 5.2 billion passenger journeys recorded in 2008/09, before the recession began, was the highest figure for 20 years.

With the onset of the recession, the growth trend was reversed once more: demand in 2018/19 was 7.9% down on 2008/09, with the figure outside London being 13.7%. Even within those figures, there have been divergent trends, with patronage falls during and after the recession being particularly strong in Scotland and Wales, whilst other areas have seen continuing growth. These are discussed in more detail in the sections on the markets for individual regions and nations which follow later in this volume.

The figures for recent trends are shown in Table 2. The figures are indexed and illustrated graphically at Figure 1-6 (for England) and Figure 1-7 (for Scotland, Wales and Northern Ireland).

Table 2: Changes in Bus Patronage (millions) since 2004/05, by Area

Year to 31 March	London	PTE Areas	English Shires	Scotland	Wales	All excl London	All GB	Northern Ireland	All UK
2005	1,802	1,069	1,177	460	123	2,829	4,631	65.1	4,696
2006	1,881	1,070	1,184	466	120	2,840	4,721	67.2	4,788
2007	1,993	1,073	1,253	476	119	2,921	4,914	67.5	4,982
2008	2,160	1,099	1,297	488	121	3,005	5,165	69.9	5,235
2009	2,228	1,105	1,330	484	125	3,044	5,272	70.5	5,343
2010	2,238	1,086	1,315	459	116	2,976	5,214	68.2	5,282
2011	2,269	1,070	1,317	431	116	2,934	5,203	66.6	5,270
2012	2,324	1,041	1,314	439	116	2,910	5,234	66.5	5,301
2013	2,311	999	1,280	423	109	2,810	5,121	66.9	5,188
2014	2,361	1,013	1,300	424	107	2,844	5,206	66.9	5,272
2015	2,364	998	1,286	414	101	2,799	5,163	66.6	5,229
2016	2,293	971	1,266	409	100	2,747	5,039	65.2	5,104
2017	2,240	938	1,260	393	100	2,691	4,931	65.7	4,997
2018	2,226	919	1,232	379	96	2,626	4,852	66.1	4,918
2019	2,199	904	1,213	377	103	2,598	4,797	68.7	4,866
% changes									
Since 04/05	22.0%	-13.7%	3.1%	-17.9%	-16.2%	-7.4%	4.1%	5.5%	4.1%
Ten Years	-1.7%	-14.9%	-7.6%	-17.6%	-11.9%	-12.0%	-7.5%	0.7%	-7.4%
Five Years	-6.9%	-7.3%	-5.8%	-9.0%	1.8%	-6.5%	-6.7%	3.2%	-6.6%
One Year	-1.2%	-1.6%	-1.5%	-0.6%	6.9%	-1.1%	-1.1%	3.9%	-1.1%

Source: Statistics from DfT and Devolved Governments.

Figure 1-6: Index of Bus Patronage in England, since 2005

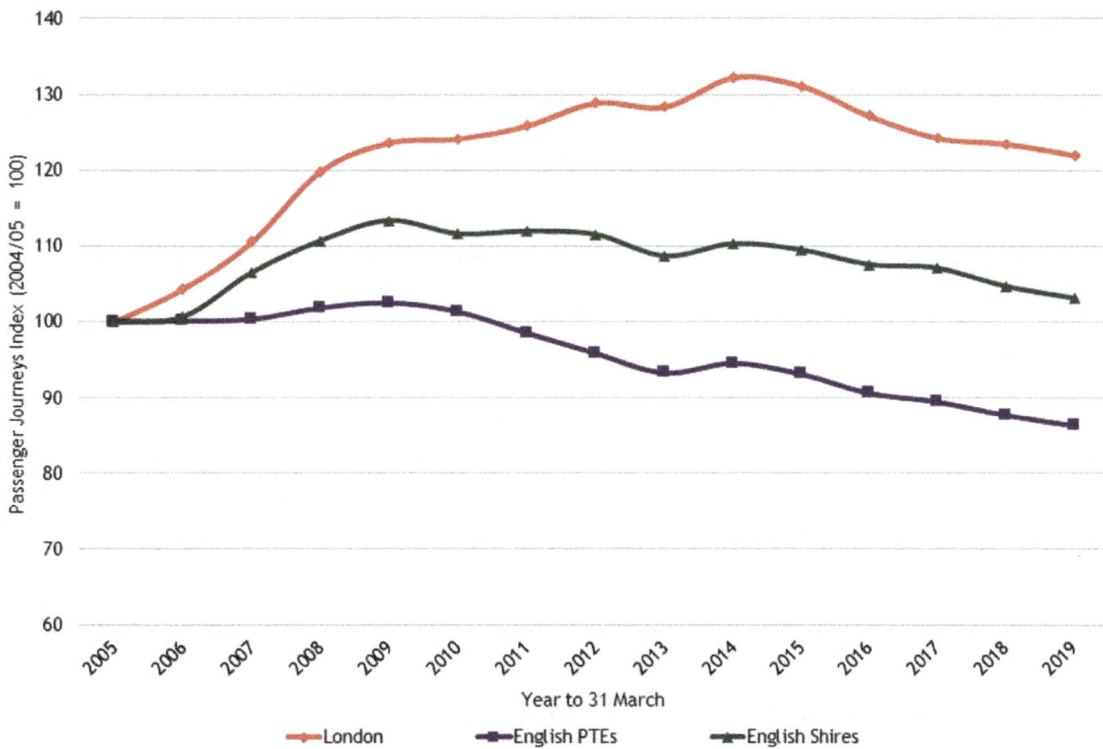

Passenger Journeys Index (2004/05 = 100)

Year to 31 March

●— London ■— English PTEs ★— English Shires

Figure 1-7: Index of Bus Patronage in the Celtic Fringe since 2005
Scotland, Wales and Northern Ireland

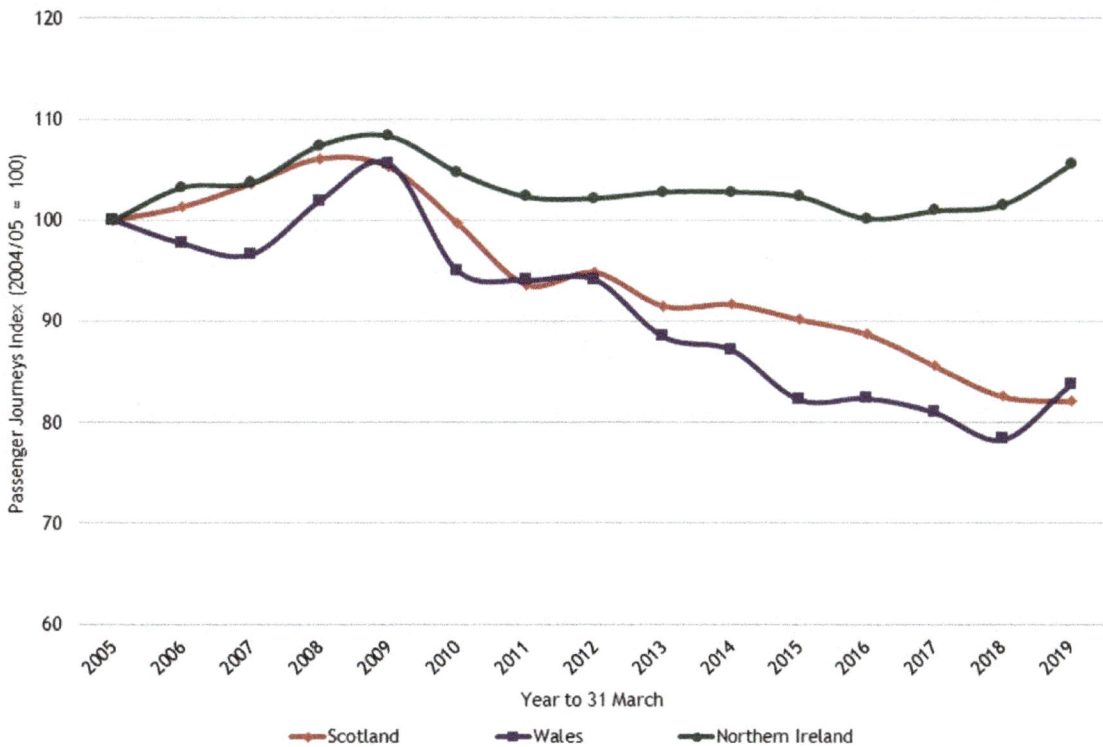

Passenger Journeys Index (2004/05 = 100)

Year to 31 March

●— Scotland ■— Wales ★— Northern Ireland

Chapter 2: The Marketing Mix for Buses

2.1 Introduction

As well as the jigsaw we looked at in Figure 1-1, there are other ways of looking at the factors which affect demand. It is useful to categorise them into two broad groups:

- **External factors** determining the competitive, economic, demographic and regulatory environment in which services are provided

- **Internal factors** which affect the product 'design' including such items as price, frequency, quality and reliability.

The various relationships are illustrated in diagram at Figure 2-1, an adaptation of the classic "marketing mix" illustration. Here, the external factors are outside the circles, whilst the internal ones are contained within them. This is a useful approach, since it makes it clear which factors the industry itself is capable of reforming or amending, and those which need outside help.

One unique feature of the industry, though, is that it is not even in full control of its own product, since it has to share highway space with other traffic, and to rely on local authorities for other key infrastructure such as stops, shelters and terminals.

Figure 2-1: The Marketing Mix for the Bus Industry

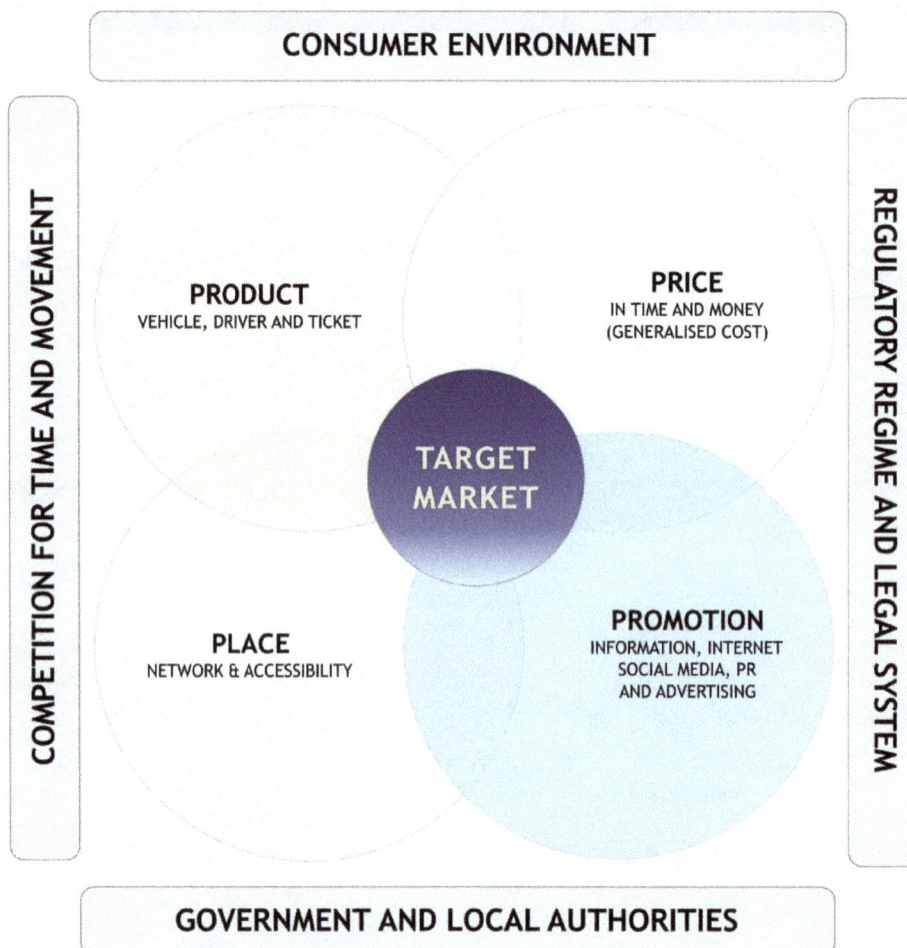

CONSUMER ENVIRONMENT

COMPETITION FOR TIME AND MOVEMENT

REGULATORY REGIME AND LEGAL SYSTEM

PRODUCT
VEHICLE, DRIVER AND TICKET

PRICE
IN TIME AND MONEY
(GENERALISED COST)

TARGET MARKET

PLACE
NETWORK & ACCESSIBILITY

PROMOTION
INFORMATION, INTERNET
SOCIAL MEDIA, PR
AND ADVERTISING

GOVERNMENT AND LOCAL AUTHORITIES

In this section of the report, we consider each of these in turn, and go on to report on key statistical trends, showing how they impact on a series of distinct geographical markets – though, as we shall see, circumstances (and therefore performance) vary quite widely even within quite tight geographical areas.

2.2   External Influences

2.2.1   The Consumer Environment

This term covers a wide variety of issues concerning the number, distribution, prosperity and tastes of consumers.

In transport, the nature of demand is special: generally speaking, movement is not a product that is desired for itself, like foodstuffs or a computer. Rather, it is a means to an end – the 'end' might be getting to work or a shopping trip or a visit to friends or relatives. Thus, demand for travel is **derived** from consumers' desire to do other things.

It follows that, in a given locality, several purely statistical factors will determine the nature and extent of people's travel. This will include things about the population, such as the number of people in a given area (population density) and their ages and their gender (this will determine the relative proportions of journeys for different purposes, such as work, school/college, shopping and leisure). These issues are further discussed in Chapter 4 below.

Thus, the choices people make about whether to travel will be influenced by a whole range of factors, including:

- whether they are in work or not
- whether they are attending school or college
- what they do with their leisure time
- how prosperous they feel.

The question of the reasons for people's journeys is discussed in greater detail in Chapter 7 below.

The consumer environment also encompasses the important 'softer' issues about customer perceptions and attitudes: fashion and "the latest thing" continue to be a vitally important element in the market for all goods and services.

Thus, for example, bus services have been viewed unfavourably by many consumers for many years, which is why many managers place such emphasis on the industry's 'green' credentials – seen as a means of rehabilitation of the product amongst the concerned middle classes.

Public attitudes are also important in indirectly influencing the way the industry is viewed by politicians at local and national level, which in turn has a crucial effect on such issues as traffic priorities, overall transport policy and regulation.

2.2.2   Competition

The next element in the mix is which way consumers decide to move from home to their destination. People can walk or go by private means such as a bike or a car, or they can use public transport.

In some circumstances, consumers will also have a choice between different types of public transport (bus versus rail for example or ship versus aircraft) or between different operators on the same route.

People's **mode choice** will be determined by a whole range of factors which will be discussed later in this section – but clearly the availability of a car in a household is one of the key factors for local journeys and for longer distance domestic journeys.

There is clear evidence of a correlation between car ownership and bus use, and the growth in the number of vehicles and drivers has been one of the biggest factors in the changing market for bus services over the last 50 years.

2.2.3   Institutions and Legal Framework

These two elements have a crucial role in determining how and where a product is delivered.

It can be looked at in three ways:

- The overall governmental approach to the economy, and the institutional framework

- The approach to overall transport policy and each mode's role in that

- The industry-specific regulatory environment

Overall Approach

The wider policies of government at local, regional and national level will all impact on the way products are designed and sold.

Trade union legislation is one example where reform has had an important influence on the way products are delivered over the last 30 years.

Society's whole approach to management of the economy will have a profound effect: the nationalisation of the industry in the late 1940s, for example, was driven by a political belief in the efficacy of public ownership *per se* rather than as a specific solution for the bus industry.

The 'unwinding' of public ownership during the 1980s and 1990s was hugely controversial and bitterly resented by sections of the population: such arguments continue to influence the way policy is made.

Transport Policy

Attitudes towards transport policy also play an important part in this equation. There have been major shifts in recent years, not necessarily associated with changes of government.

An example of this was the movement away from "predict and provide" road building policies of the 1980s by the Major government, or the differing attitudes to rail privatisation by Lady Thatcher and her successor in Downing Street.

The influence of the Climate Change debate on transport policy has been profound over the last 20 years, and this is set to have a major influence in the decades to come.

Also important is the inter-action between local and central Government, and the impact of central policy guidelines on such matters as land use planning.

Industry-Specific Regulation

This is the most immediate effect of institutional and legal frameworks on any industry, as witnessed by the Competition Commission investigation into the bus industry a decade ago.

The 'deregulation' and privatisation legislation of the mid-1980s had a profound effect on the market for bus services in most of Great Britain, and the way in which the product was delivered.

Partly for historical reasons, though, the public sector retains a significantly greater role in the planning and provision of bus services in two geographical markets, London and Northern Ireland. In the latter, government retains an ownership role too, at least for the time being – and the impetus for change there in recent years has come as much from European legislation as any immediate concern to improve transport provision. With the UK's decision to leave the EU, pressure from this direction may to reduce.

2.3 Internal Factors

As with any product, the internal factors tend to receive the most management attention. This is inevitable, since these are the elements which individual managers can do most to influence.

The factors identified are also known in marketing parlance as the "four Ps" (product, place, promotion, price). Looking at the bus product:

- 'Place' would mainly cover network design. It covers such items as how accessible the route or network is to its customers and prospective customers.

- 'Product' then covers ticketing, vehicle specification and driver standards. Network Design

- 'Price' clearly refers to the fare charged. However, it can also represent the wider cost of the journey, in time as well as money (otherwise known as the "generalised cost").

2.3.2 Place

In modern parlance, this might also be called "Access". A transport service needs to take people from where they are to where they want to be, as quickly as possible and with the minimum of hassle.

There has been much debate over the years about two aspects of network design: what might be called the 'trunk and feeder' debate and the 'high frequency versus high penetration' debate.

The 'trunk and feeder' debate is about whether a network should be configured to permit end-to-end journeys by the same mode and the same vehicle, as against the (arguably more efficient) provision of very fast, frequent 'trunk' services into which other routes or operators feed – ultimately with the trunk being provided by a rail-based mode or (in particularly in Latin America) bus rapid transit systems, using high-frequency, double-articulated buses.

The related 'high frequency versus high penetration' debate is about two alternative visions of bus service provision to housing estates and densely populated suburban areas.

The 'high frequency' model is another version of the "trunk" service – providing a simple easy to understand pattern of frequent services, albeit with lower penetration of residential areas. The alternative model is a relatively complex network of services which run very close to people's houses in a residential area but at relatively low frequencies.

The two diagrams in Figure 2-2 below try to illustrate these issues. We are looking at the south east quadrant of a large town or city, where the central area is represented by the grey square at the top of the picture.

In the left-hand diagram, a relatively complex network, necessarily of low frequency, will allow through journeys without change from virtually every square on the grid to virtually every other square. At worst, a few journeys will require one change to be completed.

On the right-hand side, the same grid is served by a "hub and spoke" network – a very high frequency "orange" route serves four key interchange points (marked by the red circles), from which local feeder buses link to the other squares. This is much more efficient from an operational point of view but will oblige a far higher proportion of passengers to change vehicles in order to complete their journey.

Figure 2-2: Two Network Design Illustrations[1]

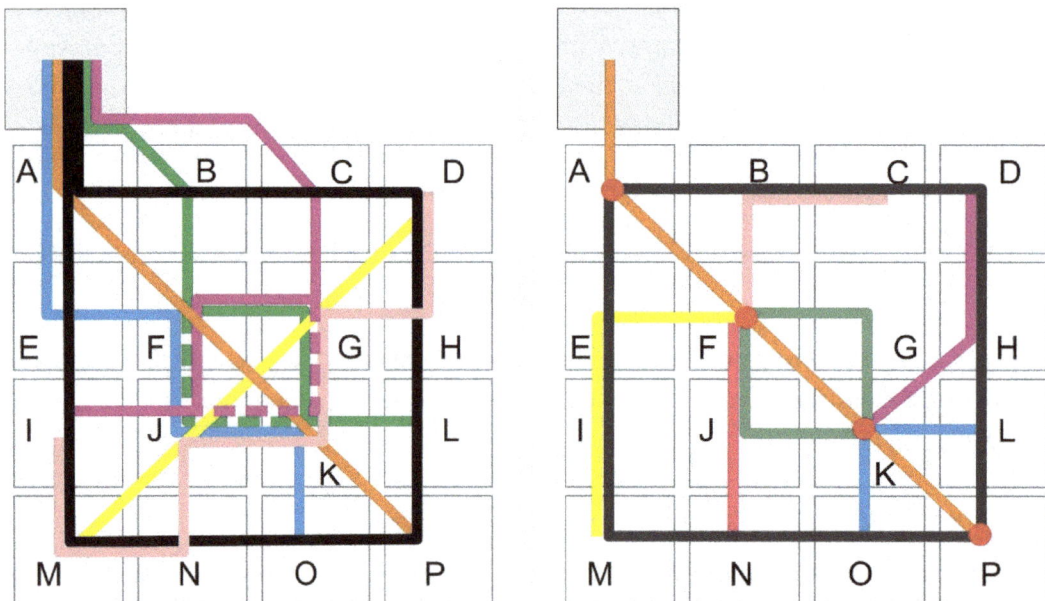

2.3.3 Product Design

Though it can seem homogenous (particularly to the non-user), there are a surprising number of variables: different elements of bus operation that can be varied according to specific circumstances, groups of customers or operating territory. Bus service provision is not – or should not be – a matter of "one size fits all".

The UK industry has become increasingly sophisticated in its approach to this issue over the last 25 years: examples include the 'cheap and cheerful' Magic Bus brand run by Stagecoach Group on a busy student corridor in Manchester, or the luxurious, spacious, leather-seated double deck vehicles famously introduced by Harrogate & District on its upmarket inter-urban 36 route linking Leeds with Harrogate and Ripon – a concept much

---

[1] *Based on illustrations given in Alan Cannell's book* Bus Rapid Transit in Latin America *(PTIS, 2008)*

imitated since, not least by the Stagecoach Gold brand originally launched in Leamington, Cheltenham and Perth, but now rolled out to routes all over the country.

Seat spacing and vehicle design are important – witness the market success of the 36, and the popularity of the low-floor vehicles introduced progressively since 1993. But they are not the only elements: ticketing systems and convenience of payment, information systems, facilities at bus stops and terminal facilities are all essential elements of designing a product that both retains existing customers and encourages new ones.

2.3.4 Service Quality

There is of course a straight link across from product and network design into service quality: operators can design the best network in the world and devise superb innovations in other elements of the product. But unless the service delivered to customers meets their aspirations in terms of speed, reliability, punctuality and customer-friendliness, the product will fail, and bus services will be relegated to a distress purchase by people who have no alternative.

For all sorts of reasons, the levels of quality delivered by the industry in the 1970s and 1980s were abysmally low. Crucially, even though radical improvements have been made since in many parts of the country, they are not yet universal, and this continues to feed low perceptions amongst non-users of services – even though customer surveys continue to show high levels of satisfaction amongst users.

Growing traffic congestion in recent years has had a particularly severe impact on this aspect in recent years, reducing operators' ability to deliver reliable, punctual and predictable services. As well as reducing market appeal, this does of course increase the costs of operation (discussed in more detail in our parallel annual *Bus Industry Performance* reports.

2.3.5 Price

Price is always a crucial element in any marketing mix, though customer research amongst bus users tends to show that concerns about fares and ticketing products tend not to be top of customers' agendas. The time taken for the journey, and its predictability, regularly show much higher levels of importance in surveys.

As with all products, a floor is given to the price operators can charge by the costs of production. Real-term costs of operation rose sharply in the run up to the recession, thanks primarily to increased labour and fuel costs. Thus, between 1997 and 2009, real-term costs of operation rose by over 36%. Since the onset of the recession labour costs have fallen back in real terms for the time being, but fuel costs continue to be volatile: they doubled in real terms between 1999 and 2011, before falling sharply back again.

At the same time, the falling market for bus services over the last 60 years has meant that operating costs per passenger have risen even more rapidly. Demand levels also have an important effect on pricing, since passenger volumes determine the levels at which a bus service can cover its cost of operation.

Thus, for example, the average load on each bus was just under 27 people in 1955. On a journey that costs £25 to operate, this meant that the revenue per passenger required to break even was 94p. By 2018/19, the average load had fallen to just 11.7, so that the revenue per passenger needed to cover the same costs more than doubles to £2.13.

More detail on the trends in fares is contained in Chapter 3 below.

2.3.6   The Time Element

As foreshadowed earlier, whereas the choice of most products in the marketplace is a function of price and quality, transport products have a third key element, and that is time. Time is important in three ways:

- It determines the cost of providing the journey

- It influences the choice that consumers make between different forms (or modes) of transport

- It is a key measure of economic efficiency, as time wasted through congestion impairs economic growth and prosperity

The Cost of Providing the Journey

The time that a bus takes to get from one end of its journey to the other has a crucial influence on the cost of operating the journey. Journey time will dictate the number of buses needed to run the service and, therefore, the size of the depot and the number of engineers needed to maintain them. The time will also dictate the number of drivers needed and so the number of managers, supervisors, payroll clerks and other support staff and equipment such as computers and ticket machines.

Also, the speed at which the bus can go during its journey will have a decisive effect on the amount of fuel consumed. This in turn influences the local air quality and the carbon emissions.

Influencing Mode Choice

Measuring the time taken to undertake a journey is a vital part of the choices that people make about how to make that journey – and, increasingly in the digital age, whether to make the journey at all. Whereas consumers will choose between most goods and services on their perception of the best balance between price and quality, this is not true for transport – three items must be balanced in mode choice decisions – price, quality *and time*.

Economic Efficiency

Efficient transport is considered key to economic efficiency. Time spent travelling is generally wasted and few people if any undertake journeys for their own sake: travelling can be stressful, particularly if it is unreliable or unpredictable and subject to congestion.

However, increasingly people now often use the time spent on a public transport journey productively, preparing for work, catching up with e-mails, etc. This gives public transport modes, including buses, an advantage over the car driver whose sole focus is (or ought to be) on driving the car rather than doing other things.

Generally, though, it is still in most people's interests to spend as little time possible on the journey. Faster and easier journeys to work will make people less stressed, ready to be more productive in the work environment; faster journeys home will give more leisure and family time.

Measuring Time

Typically, this is measured using a concept called 'generalised cost' – which represents the total cost in time and money of a door to door journey.

The generalised cost of a journey measures the total time taken from door to door, including walking, waiting and riding in the vehicle. This can then be added to the expenditure involved, to give a total cost that can be compared with the same figures for other modes.

The typical bus journey will involve four basic components:

- Walk time - from home to the bus stop to get on the bus

- Wait time - time at the stop waiting for the bus to come along

- In-vehicle time [IVT] – the time actually spent on the bus

- Walk time – from getting off the bus to the ultimate destination.

In more complex journeys, other components can also be involved, such as the time taken to change (from one bus to another or from bus to train, for example).

The basic components of a typical bus journey are illustrated in Figure 2-3 below.

Figure 2-3: The Components of Time Cost for a Bus Journey

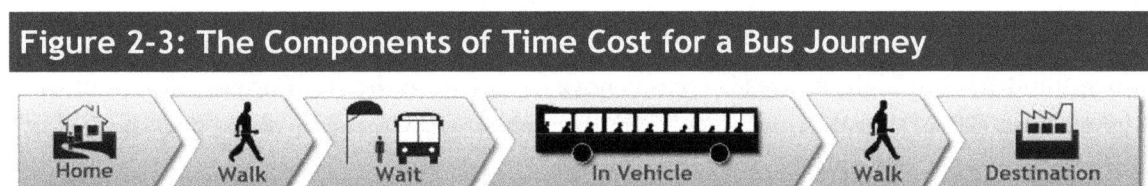

Home — Walk — Wait — In Vehicle — Walk — Destination

Chapter 3: Pricing & Revenue

3.1   Introduction

As was discussed briefly in the chapter on the marketing mix (see paragraph 2.3.5 above), the price charged for a transport service tends to be a function of the costs of operation of the journey or route concerned, and the number of people expected to travel.

Adverse trends in both costs and patronage will tend to drive prices up by more than inflation as operators strive to recover the rise in costs from fewer passengers.

This chapter reviews the available evidence on the trends in fare levels, on the changes in cost per passenger that drive them, and on the competitive position of the bus against other modes of travel.

3.2   Fare Levels

Table 3 shows fare trends by area since 2004/05, as measured by the Department for Transport's fare indices.

These feature separate results for London, the PTE areas, the English shire counties, Scotland and Wales. Trends for individual English Regions and the PTE areas are not available.

Since 2004/05, the highest fares increases have occurred in the PTE areas (39%) followed by Scotland (27.2%) and London (25.5%). Increases were lower in Wales and the English shire counties (21.8% and 21.4% respectively).

Over the last five years, the Shires have seen the highest fares increases (up by 8.5%), followed by Scotland (6.7%) and the English PTE areas (4.0%). Wales has seen the lowest increase (0.5%) whilst Mayor Sadiq Khan's fares freeze has meant a 4.4% real fall in London.

In 2018/19, the biggest fares rise was in the PTE areas (2.0%), followed by the English Shires (1.9%) and Wales (1.3%). Fares in London were 0.9% up, whilst Scotland experienced a barely perceptible 0.2% increase.

Table 3: Local Bus Fare Indices by Area (Constant Prices, 2005=100)							
Year to 31 March	London	English PTE areas	English Shire Areas	Scotland	Wales	Great Britain	England outside London
2005	100.0	100.0	100.0	100.0	100.0	100.0	100.0
2006	103.2	109.3	105.3	102.7	102.6	105.4	107.0
2007	108.7	105.9	95.1	103.8	103.9	102.9	99.7
2008	99.9	109.2	95.9	104.8	105.6	101.8	101.5
2009	108.2	123.1	102.7	114.0	113.0	111.0	111.0
2010	116.7	118.8	99.8	111.8	111.0	111.3	107.6
2011	118.4	119.9	97.8	108.3	106.6	110.8	106.8
2012	120.5	123.6	100.5	110.1	109.0	113.4	109.8
2013	122.1	123.5	102.9	111.2	112.8	115.0	111.4
2014	122.9	123.6	103.8	112.0	111.8	115.7	112.0
2015	125.1	127.0	106.6	113.5	115.4	118.4	115.1
2016	124.6	128.0	107.1	114.9	114.2	118.7	115.7
2017	126.9	136.2	116.1	123.8	119.0	125.3	124.6
2018	124.3	137.1	119.2	126.9	120.3	126.0	126.9
2019	125.5	139.8	121.4	127.2	121.8	127.8	129.3
% changes							
Since 2005/06	25.5%	39.8%	21.4%	27.2%	21.8%	27.8%	29.3%
Last five	-4.4%	4.9%	8.5%	6.7%	0.5%	2.9%	7.1%
Last Year	0.9%	2.0%	1.9%	0.2%	1.3%	1.4%	1.9%

3.3   Comparative Travel Costs

Over the last three decades, the cost of travel by bus and rail has increased in real terms, whilst the cost of motoring has also risen – the volatility in the petrol price has resulted in spikes and declines, most notably between 2010 and 2015 – the result was a marked deterioration in the competitive position of the bus.

Figure 3-1 illustrates changes in the cost of motoring, rail and bus travel since 1988, adjusted for inflation using the Consumer Prices Index (CPI). It also plots the changes against movements in average incomes.

Over the period, motoring costs have increased by around 60% in real terms, compared with 61% for public transport. At the same time, income has more than doubled. Indeed, for most of the period, growth in incomes comfortably outstripped the increases in fares, so that commuting costs accounted for a falling proportion of incomes. Once average income growth stopped or went into reverse, during and after the recession, continuing public transport fare increases started to become a burden on hard-pressed lower- and middle-income households.

Rebasing the indices to 2005 illustrates this point clearly. Real income rises flatlined from 2007, and once the recession got fully under way real incomes fell and then flatlined again. At the time of writing, the real value of average weekly earnings was still below 2008 levels.

This widening gap can be seen from 2009 onwards on the graph. The percentage changes in each measure are summarised in Table 4 below.

Figure 3-1: Real-Term Changes in Travel Costs and Income 1988-2019

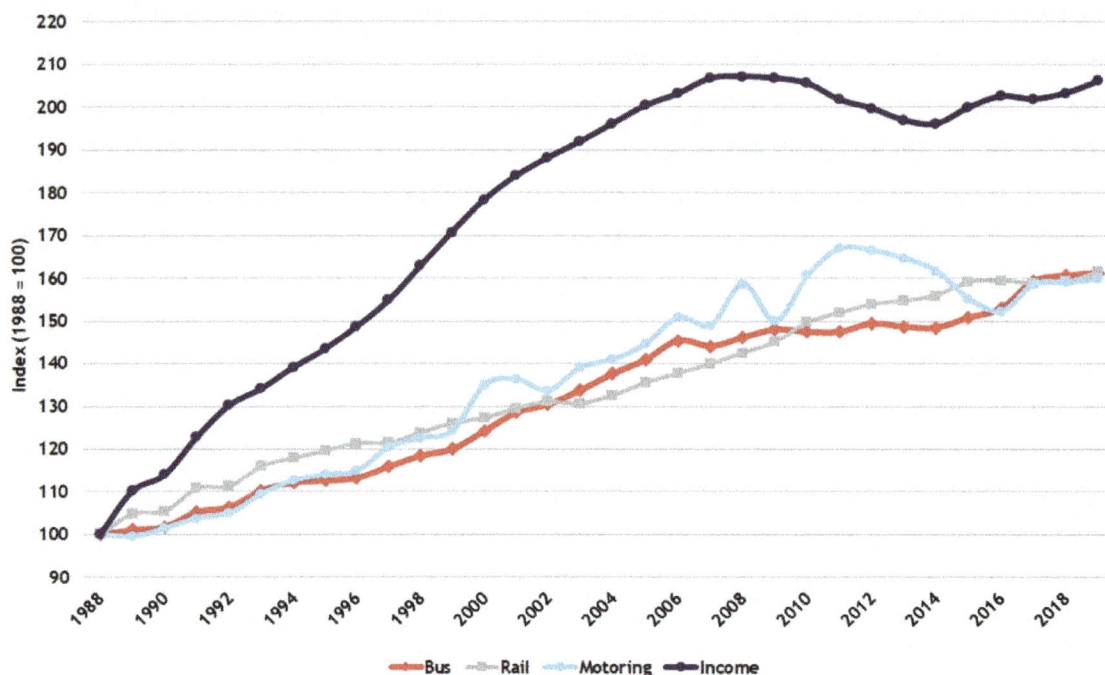

Source: Transport Statistics Great Britain/ONS.

Figure 3-2: Real-Term Changes in Travel Costs and Income since 2005

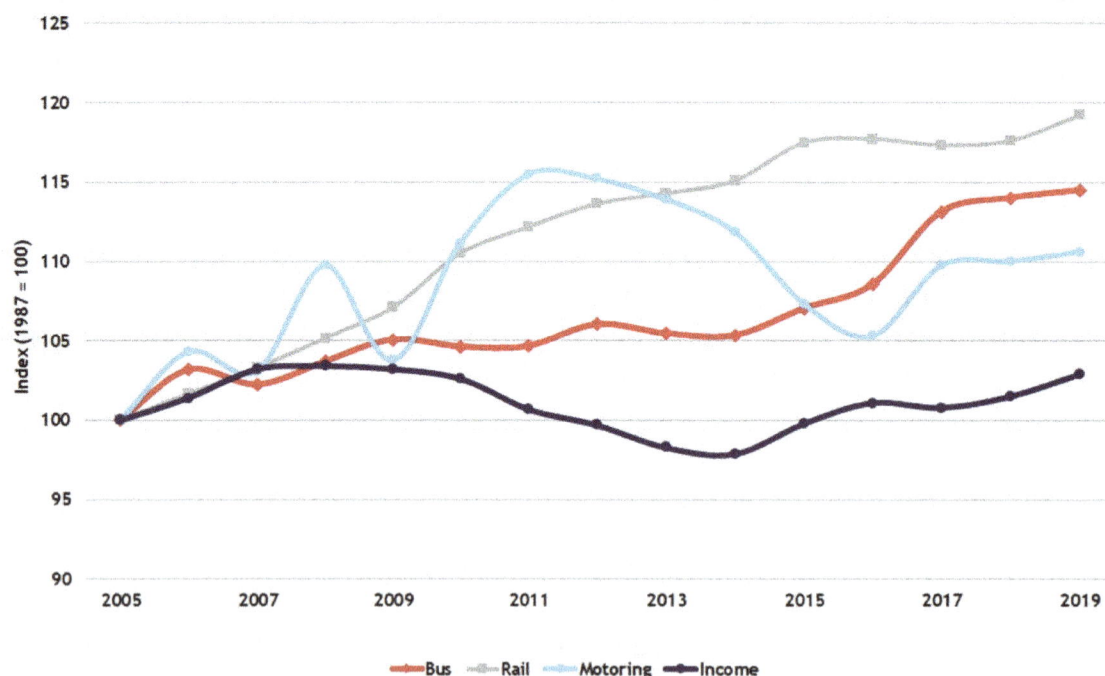

Source: Transport Statistics Great Britain/ONS.

Table 4: Changes in Passenger Transport Costs (%)

Period	Bus	Rail	Motoring	Average Weekly Earnings
Since 1988	61.3%	61.5%	59.9%	106.2%
Since 2004/05	14.5%	19.2%	10.6%	2.9%
Ten Years	9.4%	7.9%	-0.4%	0.3%
Five Years	7.0%	1.5%	3.1%	3.1%
Last Year	0.4%	1.3%	0.5%	1.4%

Source: PTIS Analysis of Transport Statistics Great Britain/ONS.

3.4 Trends in Bus Revenue

3.4.1 Overview

The annual totals, and the revenue figures for different market sectors since 2004/05, are shown in Table 5 below, in current prices, and in Table 6 in constant (2018/19) prices. The figures include concessionary fare reimbursement received from local and devolved governments but exclude BSOG and other local authority support.

3.4.2 Nominal Revenue Changes

Table 5: Passenger Revenue by Area, Current Prices

Year to 31 March	London	PTE Areas	English Shires	Scotland	Wales	All outside London	All Great Britain
2005	865	878	1,064	364	119	2,522	3,388
2006	939	888	1,119	391	130	2,690	3,629
2007	1,002	939	1,257	448	137	2,806	3,808
2008	1,048	959	1,370	482	140	2,974	4,022
2009	1,063	1,002	1,505	516	155	3,178	4,240
2010	1,124	1,032	1,551	516	155	3,255	4,379
2011	1,251	1,046	1,589	509	158	3,297	4,548
2012	1,326	1,068	1,611	527	164	3,363	4,689
2013	1,406	1,081	1,660	526	173	3,464	4,870
2014	1,493	1,087	1,709	544	164	3,514	5,007
2015	1,526	1,080	1,747	556	161	3,534	5,059
2016	1,515	1,070	1,742	561	161	3,533	5,048
2017	1,461	1,019	1,763	586	152	3,524	4,985
2018	1,438	1,080	1,697	574	149	3,506	4,944
2019	1,442	1,102	1,696	574	165	3,545	4,988
% changes							
Since 2005	66.7%	25.6%	59.4%	57.7%	38.4%	40.5%	47.2%
Last five years	-5.4%	2.1%	-2.9%	3.3%	2.5%	0.3%	-1.4%
Last Year	0.3%	2.1%	-0.1%	0.0%	10.7%	1.1%	0.9%

Source: Statistics from DfT and Devolved Governments

As can be seen, the largest growth in cash terms over the period has been in London, which has seen an increase of over 66%. Next come the Shire Counties, on 59.4% followed by Wales (40.5%) and Scotland (38.4%). The PTE areas have seen the lowest growth in revenue, at 25.6%.

In the most recent five-year period, London's revenue has fallen by 5.4% and the English Shire areas by 2.9%. Scotland has seen the highest growth on 3.3% with Wales on 2.5%.

The most recent twelve-month period saw a cash fall in the English Shire areas (a barely noticeable 0.1% reduction), with a standstill in Scotland, and growth of 2.1% in the PTE areas. Wales saw an increase of 10.7%.

3.4.3   Real-term Revenue Changes

Looking at revenue in real terms (adjusted for inflation by the CPI), we see that London has succeeded in growing its earnings by 26.9 % since 2004/05; next come the English Shires on 21.4% and Wales on 5.3%. Revenue in Scotland has fallen by 4.1% and in the PTE areas by 4.4%.

Over the last five years, operators in all areas have seen their passenger revenue fall in real terms. London has seen the sharpest fall, of 11.6%, followed by the English Shires on 9.2% and the PTE areas on 4.6%. Companies in Wales have seen a 4.2% drop.

The real-term position in the most recent twelve-month period meant that there were falls in the value of revenue received in London (1.5%), the English Shires (1.9%) and Scotland (1.5%). There were however gains in Wales (8.7%) and the PTE Areas (0.2%).

Table 6: Passenger Revenue by Area, Constant (2018/19) Prices

Year to 31 March	London	PTE Areas	English Shires	Scotland	Wales	All outside London	All Great Britain
2005	1,137	1,153	1,398	607	156	3,314	4,451
2006	1,205	1,138	1,435	711	167	3,450	4,655
2007	1,250	1,171	1,567	590	171	3,499	4,749
2008	1,273	1,165	1,664	615	170	3,613	4,886
2009	1,259	1,187	1,783	611	183	3,764	5,023
2010	1,310	1,203	1,807	602	181	3,793	5,103
2011	1,433	1,199	1,820	577	181	3,777	5,210
2012	1,497	1,205	1,818	587	185	3,795	5,292
2013	1,554	1,195	1,836	608	191	3,830	5,384
2014	1,620	1,179	1,854	600	178	3,811	5,431
2015	1,632	1,155	1,869	585	172	3,781	5,413
2016	1,607	1,135	1,848	594	171	3,747	5,354
2017	1,514	1,056	1,826	612	158	3,651	5,165
2018	1,465	1,100	1,730	591	152	3,573	5,037
2019	1,442	1,102	1,696	582	165	3,545	4,988
% changes							
Since 2005	26.9%	-4.4%	21.4%	-4.1%	5.3%	7.0%	12.1%
Last five years	-11.6%	-4.6%	-9.2%	-0.5%	-4.2%	-6.2%	-7.9%
Last Year	-1.5%	0.2%	-1.9%	-1.5%	8.7%	-1.5%	-1.0%

3.4.4 Indexation

Another way of looking at the changes in revenue is to look at an index, taking the 2004/05 figure as equalling 100. This allows the historic plot back to deregulation to be made, whilst taking into account the break in time series caused by the DfT's methodology changes.

The results of these calculations are shown in Table 7 below, and illustrated graphically at Figure 3-3 for England and Figure 3-4 for Scotland and Wales.

Figure 3-3: Bus Revenue Index, England, 2018/19 prices

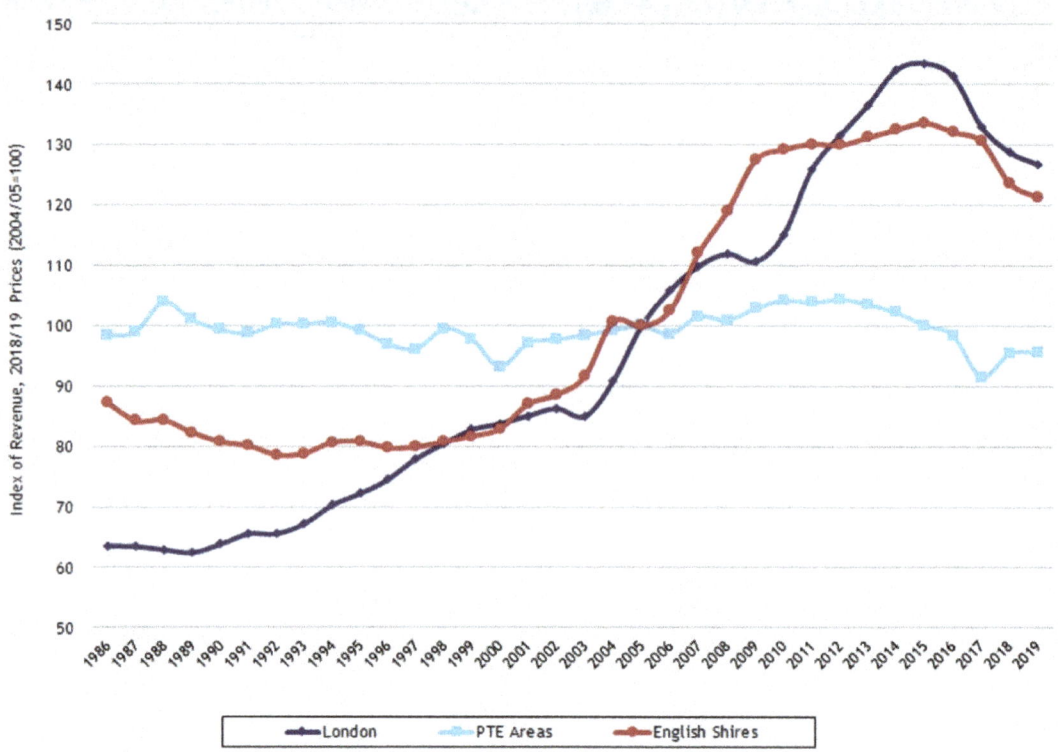

Figure 3-3: Bus Revenue Index, England, 2018/19 prices

Legend: London, PTE Areas, English Shires

Figure 3-4: Bus Revenue Index, Scotland, Wales and GB 2018/19 prices

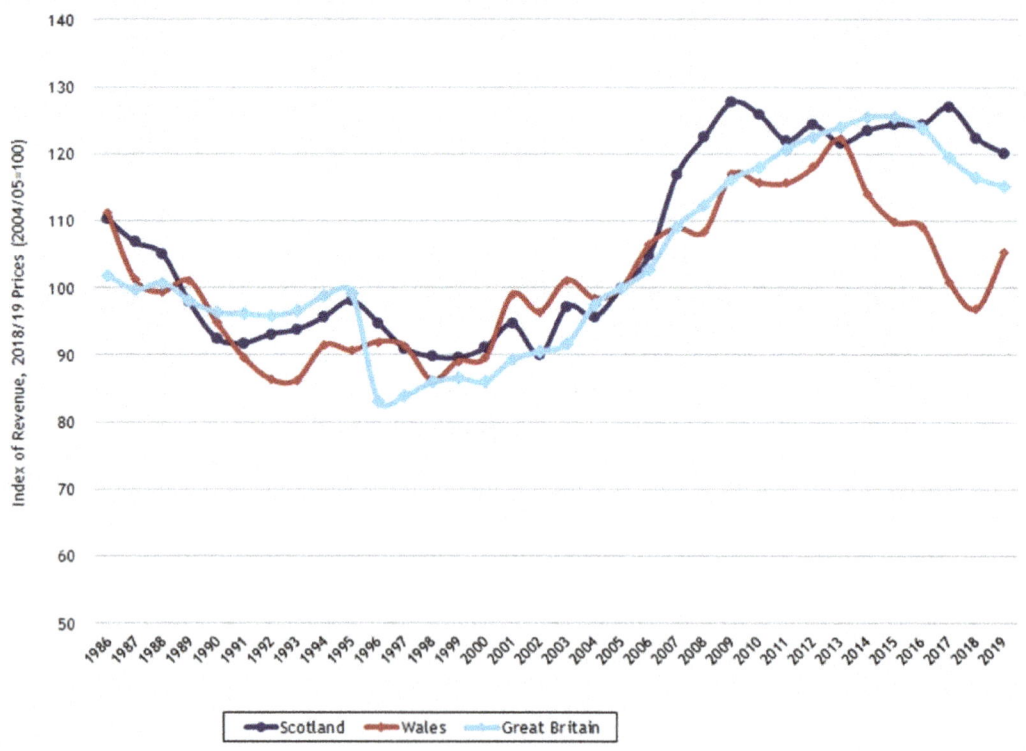

Figure 3-4: Bus Revenue Index, Scotland, Wales and GB 2018/19 prices

Legend: Scotland, Wales, Great Britain

The process of taking the analysis back to 1986 produces some interesting results. It is particularly noteworthy that, in real terms, the value of passenger revenue in Wales and the PTE areas was lower in 2016 than in 1986. The fall in both Wales and the PTE areas is particularly substantial, especially given the significant real increases in operating costs that have occurred over the last 20 years.

Table 7: Index of Bus Revenue, 2018/19 prices
2004/05 = 100

Year to 31 March	London	PTE Areas	English Shires	Scotland	Wales
1986	63.56	98.38	87.33	110.25	111.27
1987	63.48	99.00	84.34	106.83	101.18
1988	62.96	104.01	84.47	105.07	99.39
1989	62.49	101.05	82.41	97.82	101.04
1990	63.88	99.37	80.85	92.44	94.84
1991	65.59	98.76	80.18	91.76	89.59
1992	65.59	100.24	78.65	93.08	86.35
1993	67.21	100.27	78.87	93.74	86.24
1994	70.34	100.57	80.74	95.64	91.42
1995	72.23	99.16	80.86	98.06	90.68
1996	74.55	96.97	79.81	94.78	91.94
1997	77.97	96.16	79.97	90.95	91.34
1998	80.49	99.47	80.89	89.75	86.19
1999	82.85	97.83	81.77	89.59	89.08
2000	83.81	93.17	83.01	91.08	89.57
2001	85.14	97.15	87.11	94.63	99.02
2002	86.34	97.72	88.65	89.99	96.41
2003	85.08	98.47	91.76	97.20	101.17
2004	90.83	99.33	100.75	95.63	98.42
2005	100.00	100.00	100.00	100.00	100.00
2006	105.96	98.72	102.65	104.92	106.52
2007	109.95	101.60	112.15	116.87	109.03
2008	111.99	100.99	119.05	122.50	108.45
2009	110.70	102.95	127.60	127.74	117.01
2010	115.18	104.33	129.30	125.91	115.80
2011	126.05	103.95	130.25	121.99	115.67
2012	131.65	104.50	130.07	124.30	118.13
2013	136.73	103.65	131.35	121.59	122.27
2014	142.50	102.27	132.63	123.52	114.04
2015	143.55	100.17	133.71	124.40	109.89
2016	141.35	98.43	132.23	124.39	109.19
2017	133.15	91.54	130.69	127.08	100.82
2018	128.85	95.43	123.76	122.35	96.91
2019	126.87	95.61	121.36	120.08	105.31

All areas have seen a reduction since a peak reached in the first half of the most recent decade with substantial falls since, as can be seen from Table 8 below.

Table 8: Recent falls in Bus Revenue – Peak Year and % fall since

Area	Year Peaked	Peak Figure	% fall since
London	2014	143.55	-11.6%
PTE Areas	2012	104.50	-8.5%
English Shires	2015	133.71	-9.2%
Scotland	2009	127.74	-6.0%
Wales	2013	122.27	-13.9%
Great Britain	2015	125.46	-8.2%

3.5   Revenue Analysis

Whilst passenger and revenue numbers give a crude indication of the general trend in demand, such statistics do not give any 'feel' for changes to the length of journeys. Nor do they show the effects of schemes such as Oyster and Travelcard in London and the spread of daily and weekly tickets. These have all had an impact on passenger numbers through a policy of marginal pricing.

We can analyse such trends with other measures, including changes in average fare paid and by comparing the changes in receipts with the income that would be expected as a result of headline alterations to bus fares.

3.5.1   Average Fares

The trends in revenue per passenger journey (average fare paid) are shown in Table 9 below. As well as variations in charging levels, the variations also reflect other market differences, most notably in the length of each journey. Journeys tend to be longer on rural, inter-urban and suburban services than they are in London or the other major conurbations.

London

The table shows that the figure for London in 2018/19 was 65.6p: in real terms, this is roughly at the same level as it was in 1987. However, there is a clear divide marked by a change of Mayor. The fare paid fell consistently during Ken Livingstone's period in office, reaching a low point of 57.6p in the last fiscal year covered by his decisions on fares, 2008/09. Following Boris Johnson's election in April 2008, his first decision on fares which took effect in January 2009, average fares began to rise again, reaching a peak of 7.01p in 2015/16, a rise of 24% during the Johnson mayoralty.

Sadiq Kahn famously won the 2016 election on the back of a promised fares freeze, and this is evidenced in the 6.4% fall in average fare since.

England outside London

By comparison, passengers in a PTE area pay on average £1.21 each, 2.6% higher in real terms than five years ago, and 7.2% up over the last decade.

In the English Shire areas, the figure is £1.40, 3.6% lower in real terms than a five years ago. The figure reached a peak of £1.46 in 2015/16 and is now 4.2% lower.

Scotland and Wales

In Scotland, the average revenue per passenger is £1.51, up by 5.3% in real terms over the last five years. The figure reached a peak in 2016/17, standing at £1.55 but has fallen back by 2.2% since.

In Wales, the figure was £1.617, 4.9% lower than five years ago when the figure was at £1.70. Similarly, there was a peak level reached in 2012/13 of £1.759, followed by a sharp reduction to £1.655 in 2013/14.

Table 9: Passenger Revenue per Passenger Journey by Area
2018/19 Prices (CPI)

Year to 31 March	London	PTE Areas	English Shires	Scotland	Wales	England outside London	All GB
2005	0.631	1.101	1.188	1.041	1.272	0.909	0.938
2006	0.640	1.085	1.212	1.078	1.388	0.913	0.946
2007	0.627	1.114	1.251	1.174	1.436	0.945	0.964
2008	0.589	1.091	1.278	1.202	1.354	0.948	0.944
2009	0.565	1.105	1.337	1.263	1.410	0.983	0.957
2010	0.585	1.133	1.376	1.314	1.550	1.020	0.983
2011	0.632	1.161	1.382	1.355	1.564	1.043	1.010
2012	0.644	1.200	1.386	1.364	1.597	1.054	1.021
2013	0.672	1.223	1.436	1.383	1.759	1.088	1.051
2014	0.680	1.190	1.428	1.402	1.665	1.077	1.042
2015	0.691	1.184	1.451	1.435	1.700	1.088	1.054
2016	0.701	1.195	1.460	1.460	1.686	1.095	1.068
2017	0.676	1.126	1.449	1.545	1.583	1.071	1.047
2018	0.658	1.197	1.404	1.542	1.574	1.078	1.037
2019	0.656	1.214	1.399	1.511	1.617	1.075	1.037
% changes							
Period	4.0%	10.3%	17.8%	45.2%	27.1%	18.3%	10.6%
Last Five	-4.9%	2.6%	-3.6%	5.3%	-4.9%	-1.2%	-1.6%
Last Year	-0.3%	1.4%	-0.4%	-2.0%	2.8%	-0.2%	0.1%

3.5.2   Revenue per Passenger Kilometre

This measure tends to be a more reliable comparator, since it breaks down the revenue over a common unit of distance travelled. It is also referred to as "yield per passenger kilometre".

DfT provides estimates of passenger kilometres travelled by bus (relating journeys to average lengths revealed by the National Travel Survey). This enables some calculations to be made. The figures by area since 2004/05, adjusted for inflation using the Consumer Prices Index, are shown in Table 10 below.

In many ways, the figures mirrored the trends already analysed in looking at revenue per passenger journey: we see the falls in yields in London until Boris Johnson's election in

2008, and the volatility in the rest of the country as first the recession and then reductions in local government spending hit yields.

The notable thing is how earnings per passenger kilometre have converged over the last decade across the areas outside London. Thus, in 2004/05, operators in the PTE areas were earning 20.5p per passenger kilometre in today's money, whilst those in the Shires were earning 18.5p (9.4% less), in Wales 16.6p and in Scotland just 15.6p (23.9% less).

By contrast, in 2015/16, PTE yields had fallen to 19.99p, but the Shires had risen to 19.2p (just 3.7% behind), and Scotland to 19.4p (2.8%). The gap in Wales remained somewhat larger at 9.4% as yields reached 18.0p in 2015/16.

Table 10: Passenger Revenue per Passenger Kilometre by Area 2018/19 Prices (CPI)						
Year to 31 March	London	PTE Areas	English Shires	Scotland	Wales	Great Britain
2005	0.146	0.199	0.177	0.148	0.158	0.168
2006	0.143	0.199	0.183	0.145	0.178	0.169
2007	0.143	0.201	0.186	0.158	0.182	0.172
2008	0.135	0.189	0.184	0.163	0.175	0.166
2009	0.133	0.188	0.188	0.162	0.178	0.167
2010	0.140	0.192	0.196	0.160	0.193	0.173
2011	0.153	0.202	0.200	0.168	0.180	0.181
2012	0.157	0.211	0.200	0.186	0.177	0.185
2013	0.160	0.213	0.193	0.198	0.199	0.187
2014	0.162	0.211	0.188	0.201	0.188	0.185
2015	0.164	0.208	0.193	0.205	0.192	0.187
2016	0.166	0.216	0.203	0.209	0.191	0.193
2017	0.160	0.199	0.204	0.221	0.179	0.189
2018	0.150	0.217	0.200	0.216	0.171	0.186
2019	0.151	0.213	0.186	0.216	0.183	0.182
% changes						
Period	3.5%	6.9%	5.3%	45.7%	15.5%	8.1%
Last Five	-7.8%	2.1%	-3.8%	5.3%	-4.9%	-3.0%
Last Year	1.0%	-1.7%	-7.1%	0.3%	6.7%	-2.1%

3.5.3   Market Gains and Losses

The trends in fares income are contrasted with the trends in fare levels in Table 11 and Table 12. This then allows us to determine the overall growth or shrinkage in the bus market. The figures are all calculated at constant (2018/19) prices.

This analysis suggests that, since 2004/05, the local bus market has grown by 1.1% in London, by 23.8% in the English Shires, by 6% in Scotland and 5.1% in Wales. However, it has shrunk by 18.7% in PTE areas.

This equates to average annualised market growth of 2.4% in the Shires, 0.1% in London, 0.7% in Scotland and 0.6% in Wales. The market has shrunk by 2.4% per annum in the PTE areas.

In the last five years, all areas have seen market shrinkage. The largest fall has come in the Shire areas at 16.3% (3.1% per annum), followed by Scotland on 9.0% (1.8%), the PTE areas on 9.0% (1.7%), London on 7.5% (1.5%) and Wales on 4.7% (0.9%).

Table 11: Calculation of Market Changes since 2004/05

	London	PTE Areas	English Shires	Scotland	Wales	All Great Britain
Total Revenue Growth	26.9%	3.7%	31.4%	21.6%	22.3%	24.0%
Total Fares Index Growth	25.5%	27.4%	6.1%	14.7%	16.3%	14.9%
Total Market Gain/Loss	1.1%	-18.7%	23.8%	6.0%	5.1%	7.9%
Annual Revenue Growth	2.7%	0.4%	3.1%	2.2%	2.3%	2.4%
Annual Fares Increase	2.6%	2.7%	0.7%	1.5%	1.7%	1.6%
Annual Market Gain/Loss	0.1%	-1.9%	2.4%	0.7%	0.6%	0.8%

Table 12: Calculation of Market Changes over the last five years

	London	PTE Areas	English Shires	Scotland	Wales	All Great Britain
Five Year Revenue Growth	-11.6%	-4.6%	-9.2%	-3.5%	-4.2%	-8.2%
Five Year Fares Increase	-4.4%	4.9%	8.5%	6.7%	0.5%	7.1%
Five Year Market Gain/Loss	-7.5%	-9.0%	-16.3%	-9.6%	-4.7%	-14.2%
Annual Revenue Growth	-2.2%	-0.9%	-1.8%	-0.7%	-0.8%	-1.6%
Annual Fares Increase	-0.9%	1.0%	1.6%	1.3%	0.1%	1.4%
Annual Market Gain/Loss	-1.5%	-1.7%	-3.1%	-1.8%	-0.9%	-2.7%

Chapter 4: Demographic Influences

4.1   Introduction

This chapter builds on the marketing mix description by providing some statistical analysis of the external components described in Chapter 2. The subjects covered are:

- Demographics and particularly population density

- Age and Gender

- Regional Trip Rates

- Household Income

- Socio-Economic Classification

Because much of the analysis is based on the National Travel Survey, the figures and analysis used in this chapter refer primarily to England outside London. Available data for London, Scotland and Wales is reviewed and analysed in later chapters.

4.2   Population Density

Measured by reference to the 2018 Mid-Year Population Estimates supplied by the Office for National Statistics, population density in the United Kingdom is 273 persons per square kilometre.

However, this disguises huge variations between the most densely populated area and sparsest. The London Borough of Islington was the most densely populated, accommodating 15,943 persons per square kilometre in 2018. At the other end of the scale, the Na h-Eileanan Siar (Western Isles) has a population density of 8.8.

This measure is important, since it will at least in part determine the market potential of a bus network of a given size. This is illustrated in Table 13 below, whereby we take a given area in square kilometres (110, roughly the size of several large unitary or metropolitan authorities, such as Bristol or Liverpool) and a given level of bus usage, measured in trips per person per year (we use 115, which is the average for a range of urban areas in England). We then calculate the likely level of demand in areas with different population densities.

Table 13: Bus Patronage at Different Population Densities

Population Density (persons per square kilometre)	Resulting Population Level for 110 Sq. Km	Annual Bus Passenger Journeys (Million) @ 115 per person per year
800	88,000	10.12
1,000	110,000	12.65
2,500	275,000	31.63
5,000	550,000	63.25
7,500	825,000	94.88
10,000	1,150,000	132.25

It follows from this that more densely populated areas can support a more intensive bus network, since the demand for trips generated will be much greater even if trip rates do not vary.

In the examples above, the area with the population density of 10,000, like nine of the Inner London boroughs, would generate ten times more bus trips at given trip rates than an area such as South Yorkshire, which has a population density of around 800 per square kilometre. As we will go on to discuss, trip rates will also tend to be higher in more densely populated urban areas.

The only areas of the UK with population densities of more than 10,000 are those nine Inner London boroughs. Indeed, only one area outside London even has a population density at more than 5,000 – the City of Portsmouth at 5,378.

This is another vivid illustration of the fact that London is very different from other urban areas in the UK, and that great care must be taken in thinking that lessons from London can be applied elsewhere.

Perhaps surprisingly in view of all the talk of overcrowding, other urban areas in the country are much less densely populated, as can be seen from Table 14 below. This lists the 20 local authority areas outside London with densities above 3,500 persons per square kilometre.

Table 14: UK Areas with Population Densities above 3,500

Authority	Density (Persons per Sq. Km)	Area (Sq. Km)
Portsmouth	5,378	40
Southampton	5,056	50
Luton	4,979	43
Leicester	4,866	73
Manchester	4,721	116
Watford	4,608	21
Slough	4,519	33
Liverpool	4,418	112
Nottingham	4,414	75
Southend-on-Sea	4,344	42
Birmingham	4,259	268
Bristol	4,213	110
Reading	4,080	40
Blackpool	3,980	35
Sandwell	3,807	86
Wolverhampton	3,797	69
Coventry	3,705	99
Kingston upon Hull	3,671	71
Norwich	3,619	39
Glasgow City	3,579	175
Ipswich	3,526	39

4.3 Bus Use by Age and Gender

4.3.1 Trip Rates

There is a large variation in the use made of buses by different age groups and genders. This is measured annually as part of the Department for Transport's National Travel Survey. The figures for the most recent year are shown in Figure 4-1 below.

Figure 4-1: Bus Trips by Age and Gender, England outside London, 2018

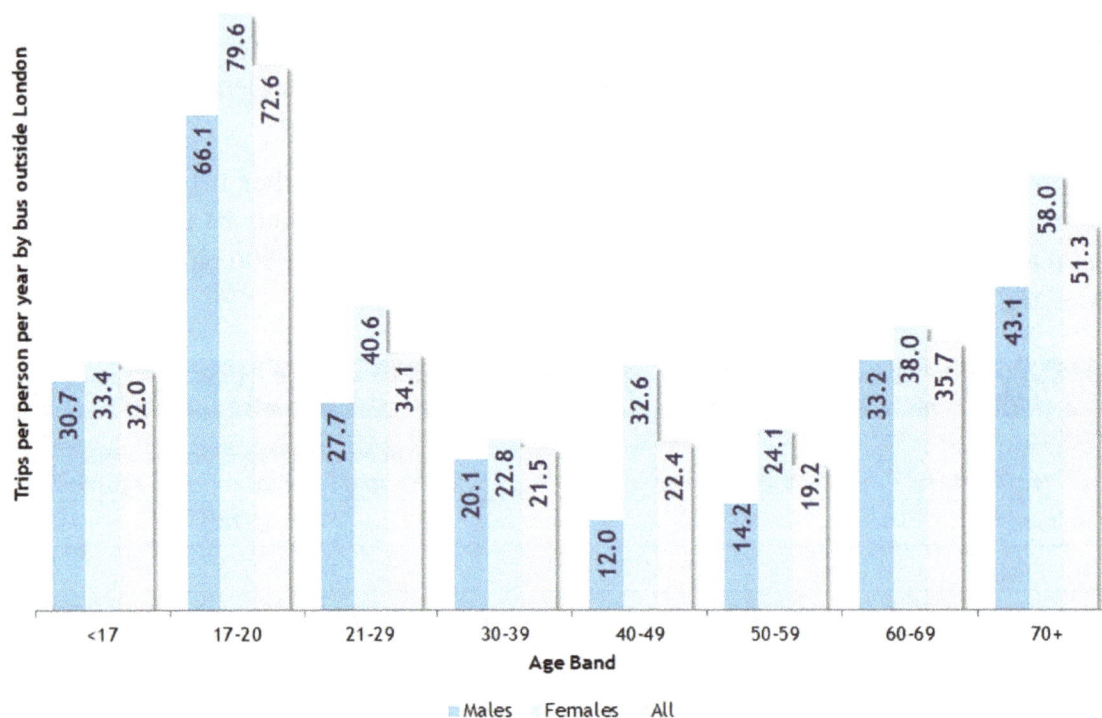

Source: National Travel Survey 2018, Department for Transport

As will be seen, women of all ages make more use of buses than men – though the difference can be exaggerated: it is not true to say that the bus product is used *overwhelmingly* by women.

In terms of age group, the biggest users of bus services are people between the ages of 17 and 20. At one stage, both males and females in this age group were making over 100 trips per person per year – though this has fallen back quite sharply in recent years.

People over 70 are the second most frequent travellers, whilst the third most important group is the 21-29 cohort. The age group with the lowest volume of trips in 2018 was people in the 40-49 and 50-59 bands with men aged between 40 and 49 making just twelve trips each year.

4.3.2 Trends over Time

The changes in trip rates over time are recorded in Table 15 below. It is apparent that there have been some major falls in trip rates over time. Across both genders, falls amongst children under 17, the two "middle age" decades of 40-49 and 50-59 have fallen by over 40%.

The largest fall has been amongst people aged 60-69, reflecting the gradual increase in the qualification age for free concessionary travel in recent years.

Table 15: Bus Trip Rates by Age & Gender, England outside London

Trips per person per year	All ages	<17	17-20	21-29	30-39	40-49	50-59	60-69	70+
Males									
2009	36.2	43.3	74.2	34.8	20.5	25.2	20.3	43.3	54.9
2010	36.2	44.2	72.3	39.0	23.0	19.0	23.9	43.9	47.6
2011	34.1	40.1	75.0	29.4	23.0	23.2	18.6	44.1	46.4
2012	35.7	39.9	67.0	47.2	24.9	18.4	23.4	34.6	53.4
2013	33.9	38.6	81.9	36.9	21.0	20.5	20.4	37.2	47.5
2014	34.2	41.1	62.1	38.5	24.6	19.9	20.8	33.2	51.5
2015	36.3	37.9	90.5	34.4	32.1	21.5	19.8	38.0	54.3
2016	30.2	36.4	57.8	35.6	21.6	22.7	11.3	30.8	43.1
2017	32.7	32.8	80.0	36.8	22.6	18.4	17.3	46.7	43.0
2018	27.7	30.7	66.1	27.7	20.1	12.0	14.2	33.2	44.5
% change	-31.0%	-41.0%	-12.3%	-25.8%	-2.0%	-109.8%	-42.8%	-30.2%	-23.2%
Females									
2009	53.9	48.5	108.3	54.1	33.5	39.1	35.4	70.4	85.2
2010	48.0	44.1	102.9	47.9	31.9	33.8	31.8	63.0	70.2
2011	49.7	42.8	121.6	49.1	33.2	35.7	40.9	58.8	70.1
2012	46.7	41.1	104.1	60.1	27.1	30.3	32.4	61.0	64.5
2013	48.9	43.2	120.4	52.6	26.1	39.9	34.3	54.8	72.2
2014	46.6	39.2	112.8	55.6	30.4	30.3	33.9	54.8	67.2
2015	46.2	44.3	92.1	60.4	29.4	20.9	34.0	52.2	72.6
2016	40.4	34.8	94.1	35.1	30.0	28.7	32.8	51.0	58.0
2017	41.9	43.0	83.6	37.1	33.8	25.3	32.7	43.2	64.4
2018	37.5	33.4	79.6	40.6	22.8	32.6	24.1	38.0	60.6
% change	-43.5%	-45.3%	-36.0%	-33.2%	-46.9%	-20.1%	-47.1%	-85.1%	-40.7%
All Persons									
2009	45.2	45.8	91.0	44.4	27.1	32.3	27.9	57.2	72.2
2010	42.2	44.2	87.4	43.4	27.5	26.5	27.9	53.7	60.4
2011	42.0	41.4	97.7	39.2	28.1	29.6	29.8	51.6	59.8
2012	41.3	40.5	84.8	53.7	26.0	24.4	28.0	48.0	59.6
2013	41.5	40.9	100.8	44.8	23.6	30.3	27.4	46.2	61.3
2014	40.5	40.2	86.9	47.2	27.5	25.2	27.4	44.3	60.2
2015	41.3	41.0	91.3	47.4	30.7	21.2	27.0	45.3	64.5
2016	35.4	35.6	75.5	35.3	25.8	25.8	22.1	41.2	51.3
2017	37.4	37.8	81.7	37.0	28.3	21.9	25.1	44.9	54.8
2018	32.7	32.0	72.6	34.1	21.5	22.4	19.2	35.7	53.3
% change	-38.4%	-43.2%	-25.3%	-30.0%	-26.2%	-44.1%	-45.5%	-60.3%	-35.5%

Source: National Travel Survey 2018, Department for Transport

4.3.3 Market Share

By multiplying the trip rates discussed above by the proportions of the population in each age band and gender, it is possible to assess the importance of each segment of the population to the current levels of bus use. This can then enable operators and authorities to target improvements and promotional activity to those segments.

Looking first at age (both genders) in Figure 4-2, we can see the importance of the under 17 market to current patronage levels - accounting for almost 20%. The next most important group are the over 70s, who account for 16.3% of trips. Next comes the 21-29 cohort, accounting for just over 14% of trips. They are followed by the over 60s (11.4%) and the 30-39 group (10.7%).

Despite their high trip rate, the importance of the 17-20 age group to the overall total is not as great as may be imagined, at 9.5%.

The spread amongst the age groups indicates that bus travel is overwhelmingly a younger person's product, with 43% of trips being made by the under 30s. This dwarfs the 30% undertaken by the over 60s. But it is clear from both the trip rates and the market share proportions that the key task – and key opportunity for the industry – is to win back those between 30 and 59.

Figure 4-2: Bus Journeys by Age, England outside London, 2018

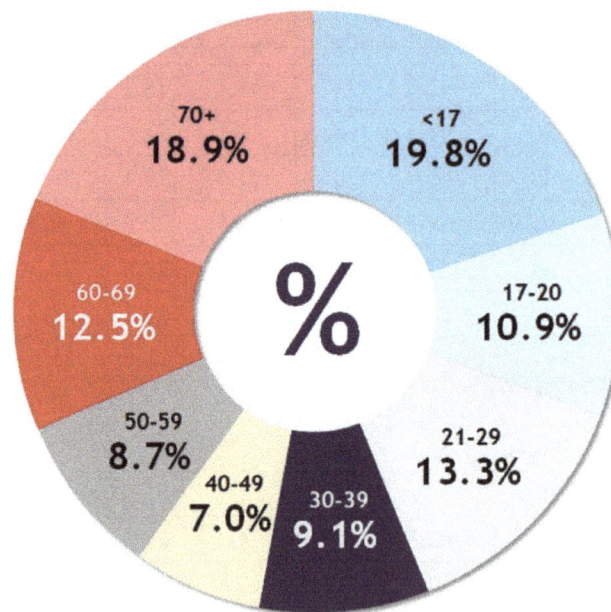

Source: PTIS, based on DfT National Travel Survey and ONS Mid-Year Population Estimates 2018

The relative importance of the gender balance can be seen in Figure 4-3 below. This takes the same estimates but looks at rather more consolidated age bands. This shows that in each of the principal age markets – 'young' (17-30), "middle" (31-59) and "elderly" (60+) – women clearly predominate.

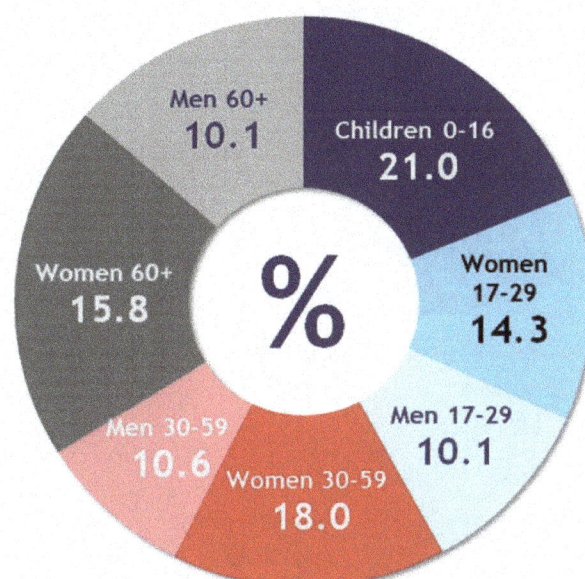

Source: PTIS, based on DfT National Travel Survey and ONS Mid-Year Population Estimates 2018

4.4 Bus Use by Household Income

The National Travel Survey also enables us to examine bus use by household income levels, as the statistics show trip rates for different income quintiles. The analysis for 2018 is shown in Table 16 below.

As can be seen, the highest volume of trip-making is, as might be expected, undertaken by the 20% of the population in the lowest income group – people in this income bracket make more than three times as many trips as those in the highest income bracket. However, the lowest income group also makes the shortest journeys.

By applying the trip rates to the total population, it is possible to estimate the relative importance of each income group to the bus market. The results are contained in the last line of Table 16, and illustrated in the graph at Figure 4-4 below.

Table 16: Bus Use in England outside London, by Income Quintile

Annual figures	Lowest real income level	Second level	Third level	Fourth level	Highest real income level	All income levels
Trips per person	52	47	27	20	16	33
Kilometres per year	374	400	253	193	159	277
Average Journey (km)	7.2	8.5	9.6	9.4	9.9	8.5
Implied % of trips	32.1%	29.0%	16.4%	12.6%	9.9%	100.0%

Source: National Travel Survey 2018 Sheet NTS0705, Department for Transport (Rows 1-3). PTIS Analysis (Row 4)

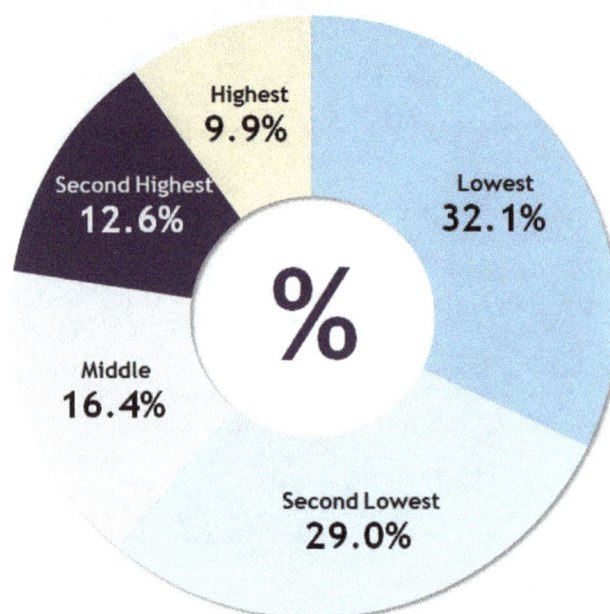

Source: PTIS, based on DfT National Travel Survey and ONS Mid-Year Population Estimates 2018

Thus, it can be said that, whilst the bulk of bus users are in the lower two income quintiles, there is still a very significant minority of users from middle- and higher-income groups. It is not true to say that buses are purely for the poorer groups in our society.

4.5   National Statistics Socio-Economic Classification

NTS data also provides a picture of the nature and extent of bus use by socio-economic classification, using the three-class version. This provides a high-level portrait, using the following classes:

- Higher managerial, administrative and professional occupations

- Intermediate occupations

- Routine and manual occupations

Other members of the population are included as either "Never worked and long-term unemployed" or "Unclassified". These are mainly full-time students.

The ONS annual population survey provides us with a percentage breakdown of the population by these groups, and the 2018 NTS provides a trip rate for each group. It is thus possible to estimate the relative importance of each socio-economic group to the market for bus travel.

The results are shown in Table 17 and the market shares are illustrated in the graph at Figure 4-5Source: ONS Annual Population Surveys (via NOMIS) Line 1. National Travel Survey Line 2. PTIS analysis Line 3.

Figure 4-5.

Table 17: Bus Use, England outside London, by Socio-Economic Group
People Aged 16-64

Item	Managerial and professional occupations	Intermediate occupations	Routine and manual occupations	Unemployed and economically inactive	Unclassified (mainly students)
% population breakdown	35.2%	28.8%	13.3%	22.8%	5.3%
2018 Trip Rate – journeys per person per year	16.3	23.7	49.4	34.6	69.7
% of trips	21.2%	25.2%	24.3%	29.2%	12.8%

Source: ONS Annual Population Surveys (via NOMIS) Line 1. National Travel Survey Line 2. PTIS analysis Line 3.

Figure 4-5: Patronage in England outside London by Socio-Economic Group
People Aged 16-64

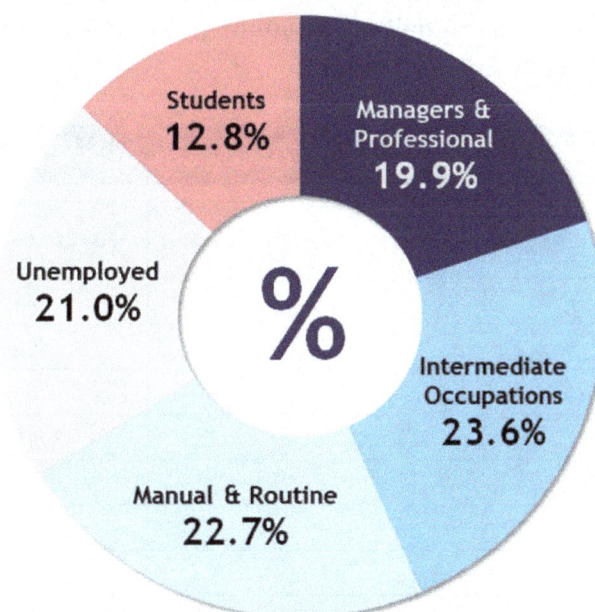

Source: PTIS, based on DfT National Travel Survey and ONS Labour Force Survey

The analysis shows that, amongst those aged 16-64, patronage is more evenly split between different occupations than might be expected. Apart from students, those in manual and routine occupations have the highest trip rate (49.4). However, there are relatively few of them these days, so that they only account for just under 23% of trips. People who are unemployed and otherwise economically inactive have the next highest trip rate at 34.6, and our estimates suggest that they account for 21% of patronage.

Those in the professional and intermediate level occupations use the bus much less frequently. However, they do now represent a very substantial proportion of the working population, so continue to be very important to the bus industry, accounting for 43.5% of trips between them. This analysis also emphasises once more the important of the student market: even though they account for just over 5% of the population, their high trip rate means that they account for almost 13% of all trips.

4.6   Regional Variations in Trip Rates

As we have seen, the overall trip rate for England outside London stood at 42 trips per person per year. However, the rate varies quite widely when broken down by region or by residential environment. This analysis combines data from two annual surveys in order to ensure a statistically significant sample in each region.

Table 18 below provides the combined figures for the most recent surveys in 2017 and 2018, but also compares them with the figures at two previous points – one in 2010/11 and another in 2014/15.

As can be seen, the North East region has the highest trip rate by some measure on 80, and indeed it increased in the most recent survey. Yorkshire and the Humber has the next highest rate on 59 – down from 64 in 2010. Next came the North West on 51, down from 69.

In the Midlands, there have been significant falls in trip rates, with the East Midlands falling as low as 28, making it the region with the lowest rate. The West Midlands has seen a hefty fall too, going from 58 in 2010/11 to 36 in 2017/18 (the latest year for which figures are available).

In regions towards the southern half of the country, trip rates have been variable but have generally fallen by a much smaller amount.

Table 18: Regional Trip Rates since 2010, bus outside London			
Region/Area	2010/11	2014/15	2017/18
North East	74	70	80
North West	69	60	51
Yorkshire and The Humber	64	60	59
East Midlands	44	42	28
West Midlands	58	53	36
East of England	30	32	29
London†	0	1	1
South East	35	38	34
South West	39	47	38
England outside London	50	48	42
England	42	41	35

† - Applies to London residents making trip on buses outside London - hence it being so low. London bus use is recorded separately and is discussed in the London chapter below.

4.7   Trip Rates by Type of Area

The Department for Transport's analysis also looks at trip rates in different types of environment, ranging from the densely populated urban areas to the very thinly populated rural villages and hamlets. The details are contained in Table 19 below.

As might be expected, the rate per person per year is much higher in the conurbations at 35, falling by more than three quarters to just 23 in the rural areas. Also notable is the sharp fall in trip rates in the conurbations since 2010, when the rate was comfortably the highest at 50. The 2017/18 rate of 35 placed the conurbations behind the smaller towns and cities whose rate was 38 – and had only fallen by a small amount since 2010/11.

Table 19: Outside London Bus Trip Rates by Urban/Rural Classification

Urban/Rural Classification	2010/11	2014/15	2017/18
Urban Conurbation	50	41	35
Urban City and Town	40	46	38
Rural Town and Fringe	36	32	28
Rural Village, Hamlet and Isolated Dwelling	20	21	23
All areas of England	42	41	35

Source: National Travel Surveys, Department for Transport. Figures combine two survey years to produce these results.

Chapter 5: Concessionary Travel

5.1 Introduction

Concessionary travel has a long history in the UK, dating back some fifty years to the mid-1960s in some areas. Free travel was introduced in London as long ago as 1973. The idea is to offer discounted or free travel on public transport modes for certain categories of people – young people, older people and people with some form of illness or disability.

By law, operators of registered local bus services are required to accept concessionary passes issued by local authorities and to give the holder a free journey or a discount. The local authority concerned is required to reimburse the operator for that journey.

However, the reimbursement paid is not the full fare that the customer would have paid. This is because the law says that the operator should not benefit financially from the existence of the scheme. This is important, since European regulations prevent concessionary travel schemes from being used to provide hidden subsidy (or state aid) to operators.

It is misleading to suggest, as some do, that concessionary fares reimbursement is a subsidy to bus operators – it is not, and great care is taken to ensure that it is not. The wording enshrined in the 1985 Transport Act is:

> "It shall be an objective (but not a duty) of an authority when formulating reimbursement arrangements to provide that operators both individually and in the aggregate are financially no better and no worse off as a result of their participation in the scheme to which the arrangements relate."

The law requires that operators should be reimbursed for:

- Revenue foregone – the fares that would have been paid by passholders who would otherwise have had to pay for their journey

- Net additional costs – extra spending operators needed to accommodate passengers using concessionary passes, including the costs of carrying the extra passengers and administrative costs incurred because of the scheme.

5.2 Changes in Entitlement to a Concessionary Pass

When concessions were first introduced in the 1960s, the determining age criterion was based on normal retirement age, which at that time was 60 for women and 65 for men. Sex discrimination legislation meant that such a distinction was illegal, so that the age had to be equalised at 60. This was done in 2002 under the Travel Concessions (Eligibility) Act 2002 and in parallel legislation in other parts of the UK.

Concessionary travel was made free in Wales in 2002. At the same time, availability was extended across the whole of Wales. A free scheme was introduced in Scotland in 2003 and the local authority schemes were replaced by a national scheme administered by Transport Scotland in April 2006.

In London, concessions are granted under separate legislation enacted originally in 1984 and updated in 1999, providing for a free scheme run jointly by the boroughs and Transport for London. This was also designed for people over 60 of both sexes.

The Department for Transport announced the English National Concessionary Travel Scheme (ENCTS) in 2007 and it was given legislative effect in the Concessionary Bus Travel Act 2007. This provided for a free pass fully inter-available across England including London for all people over the age of 60 and certain categories of disabled people.

In 2009, the government announced that the age of entitlement in England outside London would be altered to be in line with the state retirement age, and this was given effect in regulations passed shortly before the 2010 general election. Under the regulation, the eligibility changed to "in the case of a woman, her pensionable age [and] in the case of a man, the pensionable age of a woman born on the same day". Since the policies of successive governments have been to raise the female state pension age and equalise it with that of men, this means that the eligible age for the statutory concession has also increased, rising progressively from 60 in 2010 to 66 by October 2020.

This change only applied to England outside London, since responsibility for concessionary fares was a delegated one. As a result, the qualifying age in London, Scotland and Wales remained at 60. In Wales, the government announced plans in July 2019 to start the process of increasing the age of eligibility to come into line with the state pension age between 2022 and 2028. However, the idea was abandoned, and proposals removed from proposed legislation in December 2019.

Because of the changes in England, by 2019, the number of people entitled to a concessionary pass had fallen by over two million – or 20% of the population aged 60 and over. The growing gap is illustrated in the graph at Figure 5-1 below. Had the change not been introduced, the entitled population would have grown to 11.732 million in 2019. By then, we estimate that the accumulated changes had reduced the figure to 9.372 million, a reduction of 2.460 million. This has impacted on the total demand for travel by concessionary pass holders, and this is discussed below.

Figure 5-1: Increasing Qualification Age for Concessionary Passes
England Outside London

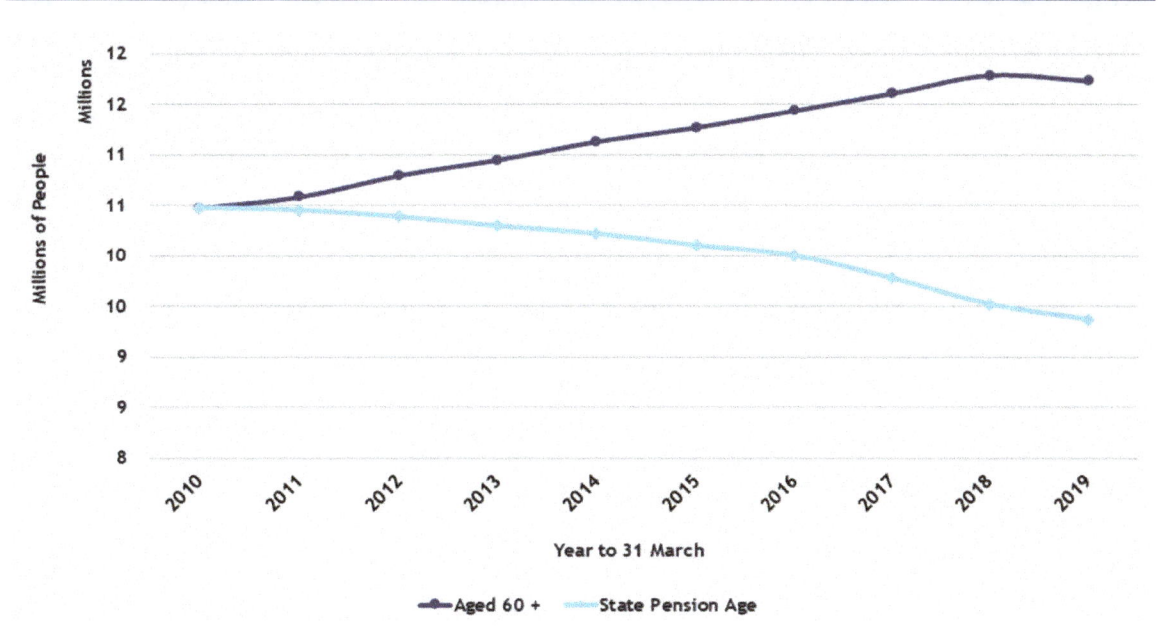

Year to 31 March

—■— Aged 60 +   —— State Pension Age

51

5.3 The Volume of Concessionary Travel

Across Great Britain, concessionary pass holders who were elderly or disabled made a total of 1,151 million bus passenger journeys in 2018/19. This was slightly ahead of the previous year, though well down from the peak of 1,236 journeys made in 2013/14. This represented 24 % of all passenger journeys. The details are contained in Table 20 below, which dates back to 2007/08, when DfT first started to publish these figures.

It will be seen that, since 2007/08, the volume of journeys undertaken in London has risen by more than a quarter but has fallen by 6.2% in Scotland and 10.7% in Wales.

In the English Shires, demand is back roughly where it was at the start of the period, having peaked in 2011/12 and fallen back since. In the PTE Areas, the peak came earlier, in 2009/10, but volumes have fallen virtually constantly since, the total reduction being 16.2% since 2008 and 14% over the last five years.

Table 20: Bus Journeys by Concessionary Pass Holders, GB since 2007/08 Elderly and Disabled People						
Year to 31 March	London	PTE Areas	Shires	Scotland	Wales	Great Britain
2008	283	273	377	151	51	1,644
2009	301	284	424	153	53	1,741
2010	318	292	441	150	49	1,775
2011	314	285	446	144	48	1,780
2012	326	287	448	147	50	1,808
2013	318	276	428	145	48	1,771
2014	335	276	429	147	48	1,802
2015	343	266	423	144	46	1,763
2016	340	254	409	142	46	1,718
2017	341	248	398	141	44	1,172
2018	351	233	378	136	45	1,143
2019	360	229	375	142	45	1,151
% changes						
Since 2007/08	27.1%	-16.2%	-0.6%	-6.2%	-10.7%	1.4%
Last five	4.8%	-14.0%	-11.4%	-1.5%	-1.1%	-5.8%
Last year	2.4%	-2.0%	-0.9%	4.6%	1.5%	0.7%

Source: Department for Transport, Annual Bus Statistics

The figures are illustrated in graph form at Figure 5-2 below, broken down into the same market segments.

Figure 5-2: Concessionary Travel Volumes, Great Britain since 2008
Elderly and Disabled People

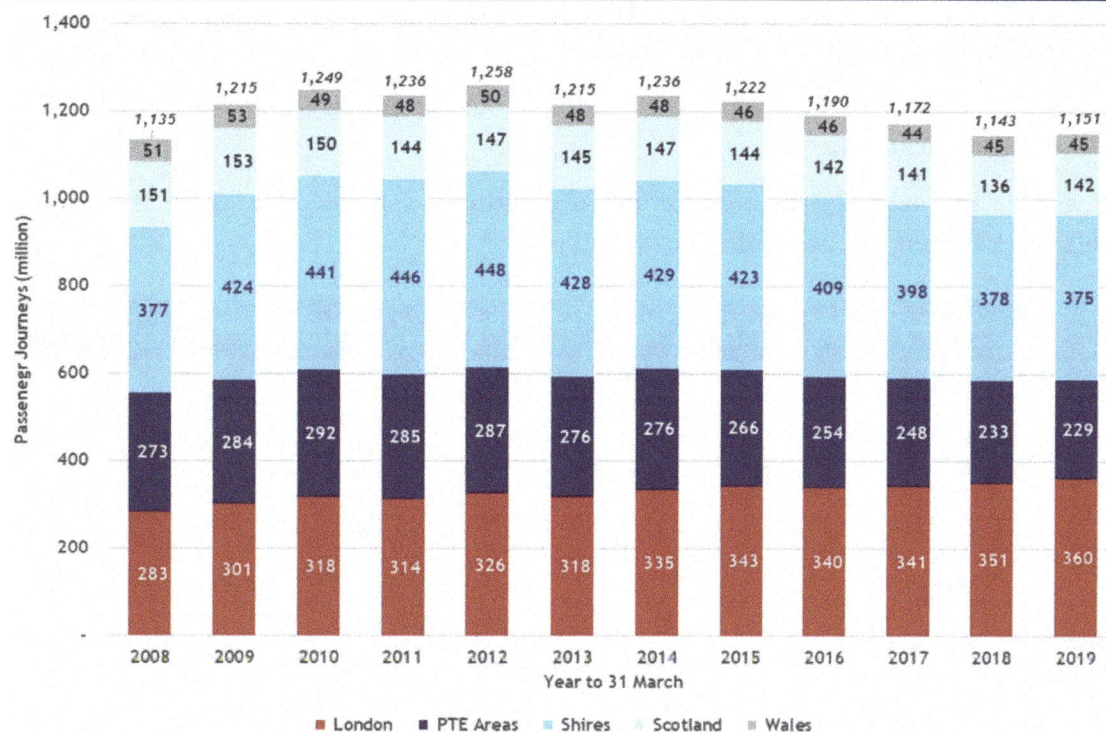

Legend: London | PTE Areas | Shires | Scotland | Wales

5.4 Changes in Car Ownership

As the current generation of older people tended to be those who were part of the huge growth in car ownership in the 1960s and 1970s, a higher proportion of today's older people are driving licence holders and therefore still likely to be driving a car, at least for some journeys. This impacts on demand for concessionary travel.

The National Travel survey tracks this annually, and the figures are shown in Table 21 below. Amongst males aged 60-69, the percentage was already high in 2009, and it has perhaps edging slightly lower. Amongst females 60-69, the proportion holding licences has tended to rise, topping 80% for the first time in 2018. A similar trend can be seen amongst the over 70s.

Table 21: Driving Licence Holders aged 60+ (%), England since 2009

Survey Year	Males 60-69	Males 70+	Females 60-69	Females 70+
2009	91.1%	76.8%	75.4%	69.1%
2010	90.0%	78.0%	77.3%	69.8%
2011	89.8%	79.0%	76.4%	70.6%
2012	90.3%	79.9%	74.9%	70.8%
2013	90.6%	81.8%	77.1%	72.8%
2014	89.8%	80.4%	77.5%	72.9%
2015	89.5%	81.2%	78.4%	73.0%
2016	89.9%	78.3%	78.6%	73.0%
2017	86.6%	80.3%	79.0%	75.4%
2018	89.9%	82.8%	83.5%	77.0%

Source: National Travel Survey, DfT

Chapter 6: Competition

6.1   Car Ownership and Use

As already noted, one of the primary influences on bus demand is the growth in ownership and use of the private car.

We can see immediately from the graph at Figure 6-1 below that there is a strong correlation between bus use and car ownership.

Figure 6-1: Car Ownership and Bus Use - Regional Comparisons

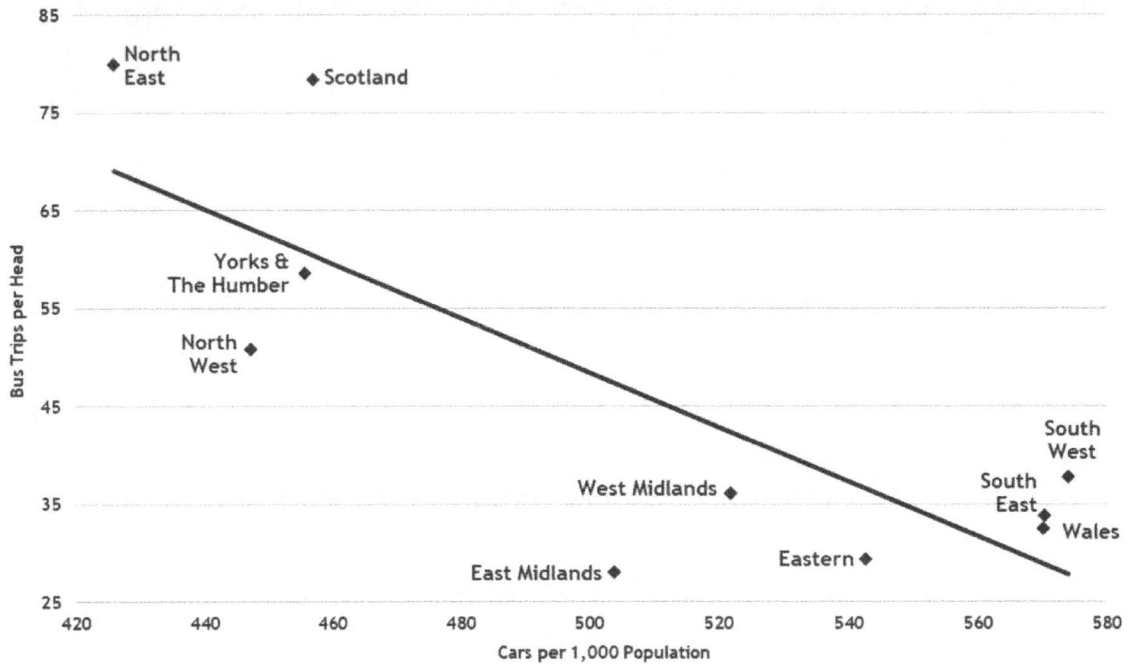

Source: National Travel Survey 2018, Vehicle Registration Statistics (both DfT), Mid-Year Population estimates (ONS)

Different levels of car ownership have a decisive impact on the volume of use for bus services, and the reason for this becomes clear when we consider data from the Government's regular National Travel Surveys. This shows that bus use by members of households without a car is three times higher than by members of car-owning households, as measured by the number of trips made per person per year. The precise difference is shown in Figure 6-2, taken from the National Travel Survey 2018.

Figure 6-2: Bus Use Outside London by Household Car Ownership

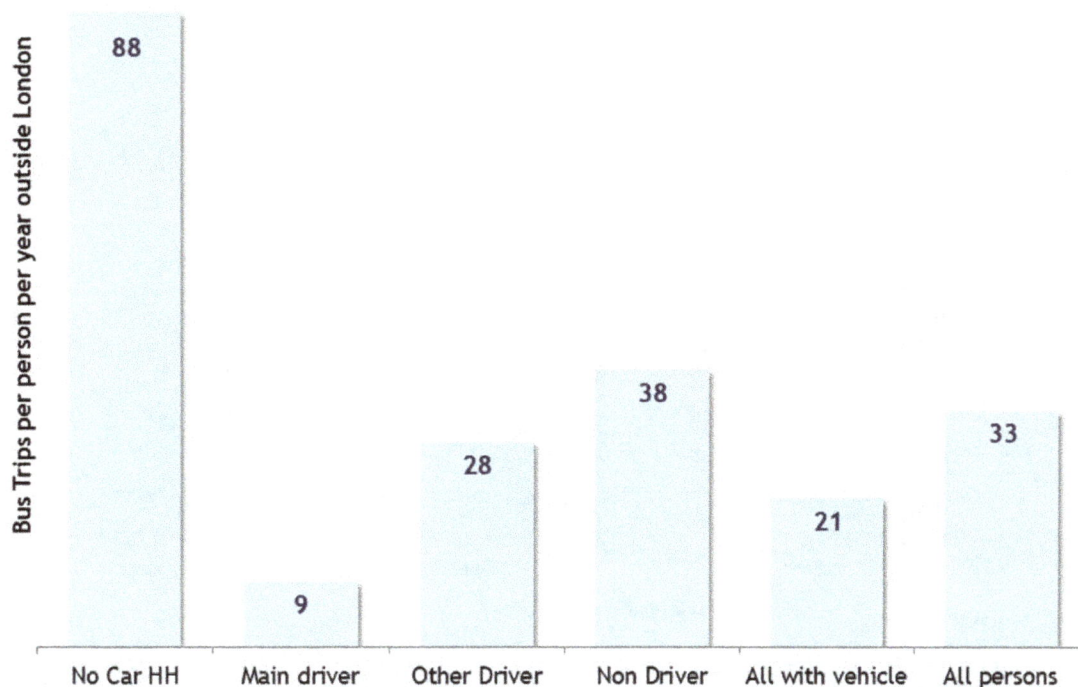

Bus Trips per person per year outside London

No Car HH	88
Main driver	9
Other Driver	28
Non Driver	38
All with vehicle	21
All persons	33

Source: DfT National Travel Survey 2018

It is therefore interesting to reflect on the changes in car ownership since the mid-1980s, and this will be discussed in more detail under the sections on geographical markets below. At this stage, it is worth noting that car ownership per thousand of the population has increased by 57% in England, 66% in Wales and a massive 88% in Scotland. Meanwhile, the same figure in London has *fallen* by 7.8%.

Outside the capital, the regions with the slowest growth have been the South East (27.9%) and East of England (44%), but then they started at a much higher level. The North East, which started with the lowest ownership figure, has experienced the biggest increase (80%).

In 1986, only the South West, the South East and the Eastern regions had more cars per person than London. However, during the 1990s, London was steadily overtaken and has had the lowest car ownership per person of any region ever since. This difference in the levels of car ownership is, in our view, a crucial factor in the market performance of bus operations in London as compared with the rest of the country over the last twenty years.

Figure 6-3: Car Ownership per 1,000 Population, by Nation

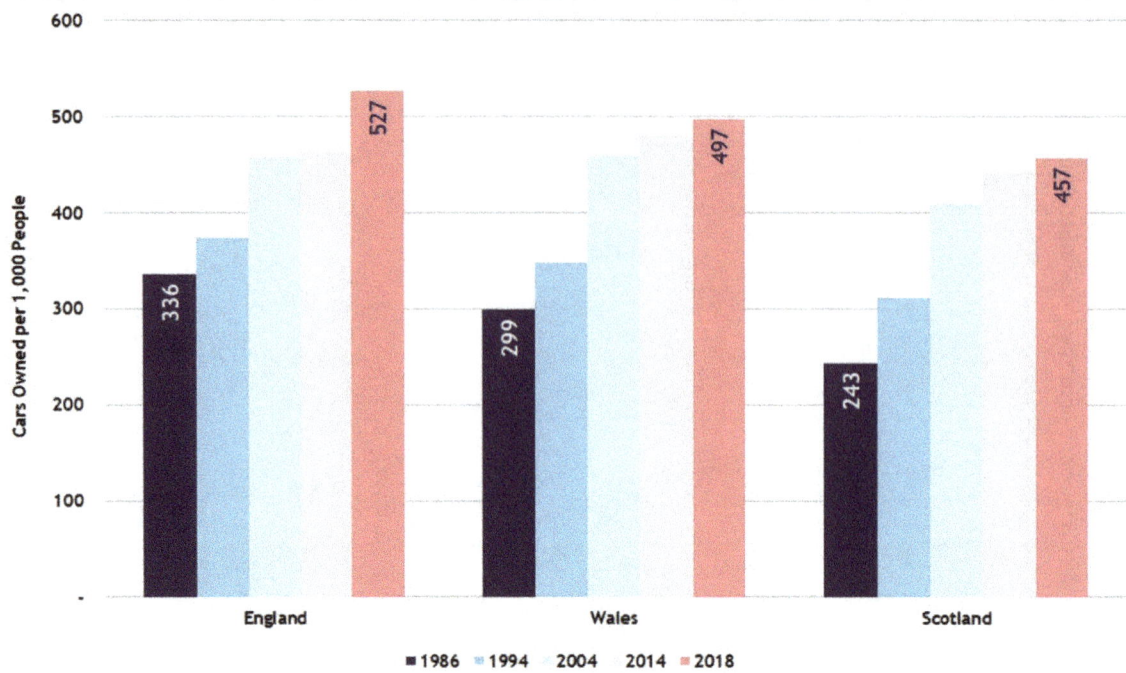

Cars Owned per 1,000 People

Nation	1986	2018
England	336	527
Wales	299	497
Scotland	243	457

■1986 ■1994 ■2004 ■2014 ■2018

Figure 6-4: Change in Car Ownership per 1,000 Population, 1986-2018

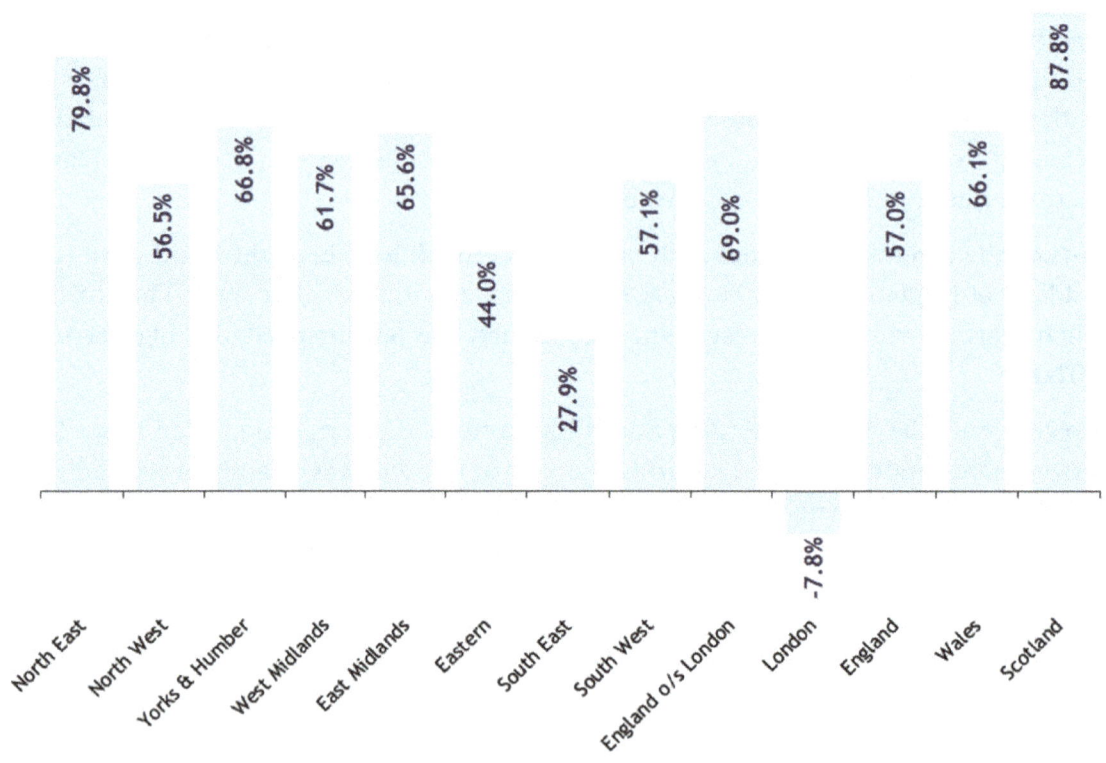

Region	%
North East	79.8%
North West	56.5%
Yorks & Humber	66.8%
West Midlands	61.7%
East Midlands	65.6%
Eastern	44.0%
South East	27.9%
South West	57.1%
England o/s London	69.0%
London	-7.8%
England	57.0%
Wales	66.1%
Scotland	87.8%

6.2 Use of Taxis and Private Hire Vehicles

6.2.1 Demand

There is little analysis of demand for travel by taxi and private hire vehicle (PHV, also known as minicabs). Regulatory authorities such as the Competition Commission and the Law Commission have previously used data from the National Travel Survey and ONS data on Consumer Trends and Family Spending to estimate the size of the market.

We have used the same approach in order to provide similar estimates – but also, importantly, to track changes in demand over the years.

Travel Surveys

Based on National Travel Survey results, it is estimated that taxis account for around one per cent of domestic local transport in England. This would imply total passenger journeys of around 546 million in 2018.

In Scotland, the Scottish Household Survey shows that taxis account for 1.4% of all journeys, compared with 8% for buses. This implies a figure for passenger journeys by taxi and PHV of around 62 million.

Unfortunately, no similar figures are available for Wales. However, it is possible to estimate a figure based on NTS data for the principality before the Welsh Government withdrew from the NTS in 2012. Relating this to current journey proportions in England suggests a figure of around 22 million passenger trips per annum.

The analysis suggests a total figure for GB of 630 million journey per annum.

Looking at these estimates over time shows that taxi and PHV demand peaked before the recession at around 648 million journeys per annum. It then dropped sharply to 589 million, before recovering to 647 million in 2016/17. However, the estimates suggest that it has fallen back again since. The trends are illustrated in Figure 6-5 below.

ONS Spending Data

Analysis of ONS data on Consumer Trends and Family Spending suggests that the spending by private consumers on taxi and chauffeur services is worth £2.9 billion, down from a peak of £3.4 billion in today's prices in the run-up to the recession. To this must be added spending by business and incoming tourists.

The figures for the years between 2005 and 2010 are actual numbers published by the ONS as part of their estimates on spending on transport by road. Since 2010, separate figures for spending on taxis and PHVs have not been published. However, figures for individual households are provided, enabling us to provide estimates of the total. These are based on the proportion of household spending on road transport that is spent on taxis.

The trends since 2005 are illustrated in the graph at Figure 6-6 below.

Figure 6-5: Domestic Taxi Demand for Great Britain, 2008-2018

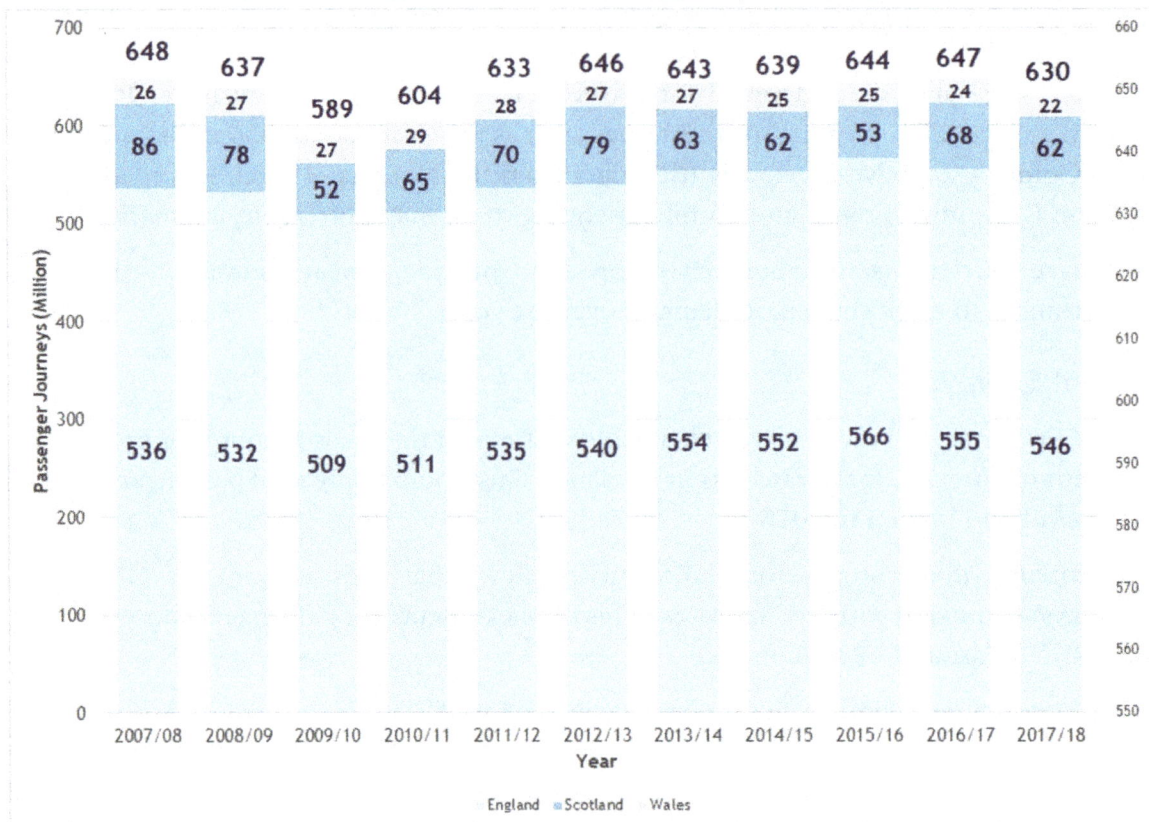

Year	England	Scotland	Wales	Total
2007/08	536	86	26	648
2008/09	532	78	27	637
2009/10	509	52	27	589
2010/11	511	65	29	604
2011/12	535	70	28	633
2012/13	540	79	27	646
2013/14	554	63	27	643
2014/15	552	62	25	639
2015/16	566	53	25	644
2016/17	555	68	24	647
2017/18	546	62	22	630

Source: Estimates based on PTIS analysis of National Travel Survey 2018 (England), Scottish Household Survey 2018 (Scotland) and PTIS estimates (Wales). Excludes incoming visitor market (see below).

Figure 6-6: Estimates of Household Spending on Taxis and PHV since 2005

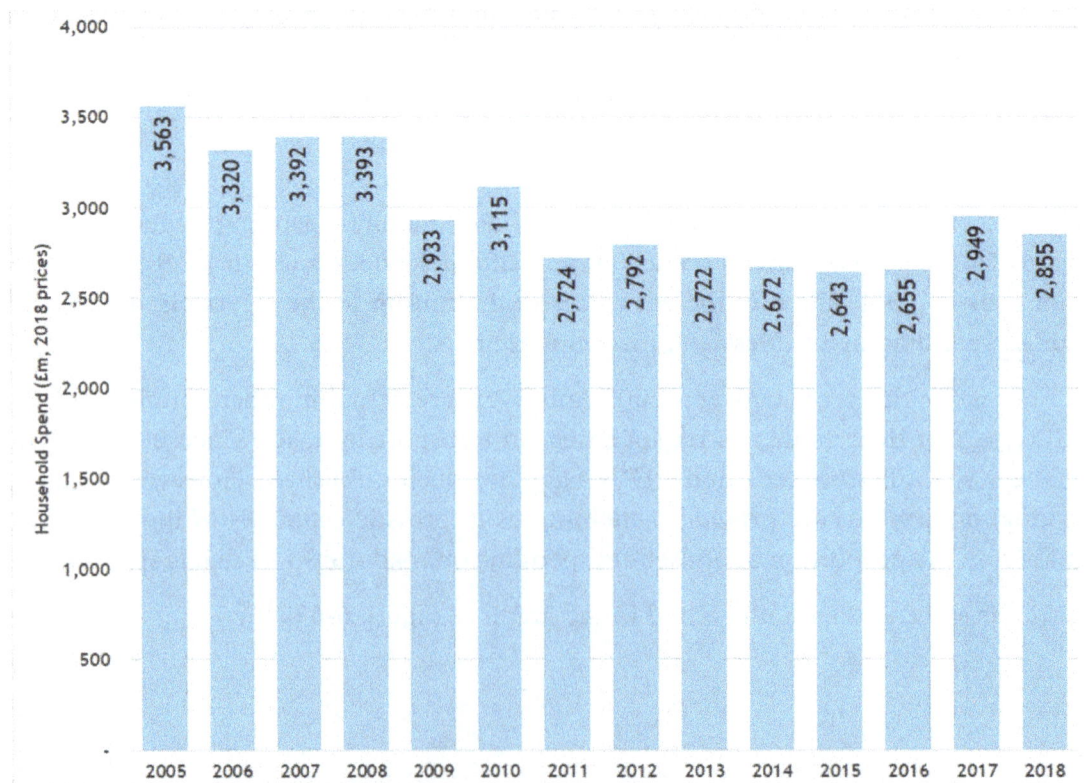

Year	Household Spend (£m, 2018 prices)
2005	3,563
2006	3,320
2007	3,392
2008	3,393
2009	2,933
2010	3,115
2011	2,724
2012	2,792
2013	2,722
2014	2,672
2015	2,643
2016	2,655
2017	2,949
2018	2,855

Source: PTIS estimates based on ONS Household Spending data

Incoming Visitors

Data on the use of taxis by incoming visitors to the UK can be deduced from the ONS International Passenger Survey (IPS) and other research carried out by Visit Britain.

In 2018, the IPS shows a total of 37.9m visits from overseas, spending an estimated 266m nights in the UK. Every year, Visit Britain adds questions about different aspects of visitor behaviour whilst in this country, including the use of internal transport services. These questions were last asked as part of the 2013 survey and the results were published in 2014[2].

This showed that 32% of visitors used a taxi during their stay – 29% of holidaymakers and 42% of business visitors. Making an assumption about the number of taxi/PHV trips per night's stay, we estimate that likely patronage levels from incoming visitors to be between 32 and 39 million per annum.

Conclusions

Looking at both the NTS data and the spending data, the market for movement by taxi and minicab does not appear to have increased over the last ten years, despite a huge increase in supply (for a discussion of which, see below).

The analysis suggests that domestic demand, at around 670 million passenger journeys in 2018, has never recovered to pre-recession levels, though it came close in 2017. It would be fair to conclude that the revolution in the business brought about by smartphone apps and ride sharing services has yet to increase the total size of the market, so that real-term average fares – at just over £5.40 per trip – are currently around 20% lower than a decade ago.

6.2.2   Supply of Services

Statistics from the DfT and The Scottish Government provide a picture of taxi and PHV provision. These record the number of vehicles and drivers and are now provided annually (formerly every two years). As will be seen, there has been a substantial increase in registration of both vehicles and drivers over recent years. Not all licence holders will be active, and many drivers may in fact be working part time.

London

Table 22 is an extract from DfT's taxi statistics and shows the number of vehicles and drivers licensed in London.

Since the mid-1980s, London has seen a 52% increase in the number of licensed taxis, with a 28% increase in the number of licensed drivers. Over the last five years, though, numbers have begun to fall again, partly no doubt driven by intensified competition from PHVs powered by smartphone apps. From a peak in 2011, there had been a fall of 7% in the number of vehicles and 5% in the number of drivers by 2018. There was a further 4.2% fall in 2019.

---

[2] *Foresight Issue 128 - Types of Transport Used while visiting Britain, published by Visit Britain research, June 2014.*

Figures for private hire vehicles are more limited, mainly because licenses were not required in London until 1998. However, the statistics published from 2005 onwards show a substantial and rapid increase in the number of vehicles, more than doubling from 40,000 to 88,100 in 2018 – driven by the growth of app-based private hire businesses such as Addison Lee, Green Tomato, Lyft and Uber. The number of licensed drivers rose even more rapidly, reaching a peak of 117,700 in 2017 – an increase of more than 200%. Interestingly, though, numbers have fallen back in 2018 and 2019, so that the 2019 total was 10.2% below the peak.

Table 22: Taxis, PHVs and Drivers since 1986 - London				
Year to 31 March	Taxis	Taxi Drivers	PHVs	PHV Drivers
1986	13.8	18.6		
1995	18.3	21.6		
1998	18.9	22.3		
2005	20.8	24.9	40.0	40.0
2007	21.6	24.6	44.4	38.0
2009	22.3	24.8	49.3	55.8
2011	22.6	25.1	50.7	61.2
2013	22.2	25.6	49.9	67.0
2015	22.5	25.2	62.8	78.7
2017	21.3	24.5	87.4	117.7
2018	21.0	23.8	87.9	113.6
2019	20.1	23.2	88.1	106.8
% changes				
Period	45.9%	24.5%	-	-
Since 2004/05	-3.0%	-7.0%	120.3%	166.9%
Since 2013	-9.3%	-9.4%	76.7%	59.4%
Last year	-4.2%	-2.8%	0.2%	-6.0%

Source: Taxi Statistics, Department for Transport

Table 23 provides the same for England outside London, Table 24 for Wales and Table 25 for Scotland.

England outside London

The number of licensed taxis in England outside London has increased by 163% since 1985/86, and by 17% since 2004/05. The number of private hire vehicles in these areas has increased at a much faster rate, increasing by 63.9%. Total licensed drivers (taxi only, PHV only and dual licences) have grown by 26% since 2004/05. As in London, the number of taxis has fallen slightly over the last two years, being 7.1% down on the 2017 peak.

Table 23: Taxis, PHVs and Drivers since 1986 – England outside London

Year to 31 March	Taxis	PHVs	Total Licensed Drivers
1986	18.9		
1995	32.9		
1998	36.5		
2000	42.1		
2002	42.6		
2004	45.9		
2005	47.3	84.5	177.2
2007	52.0	88.7	189.6
2009	53.6	101.4	192.8
2011	55.4	103.5	200.4
2013	55.8	102.7	192.7
2015	58.7	107.5	193.3
2017	59.3	122.8	213.8
2018	56.8	128.8	223.7
2019	55.4	138.5	232.7
Period	162.6%	-	-
Since 2004/05	17.2%	63.9%	26.3%
Since 2013	-0.7%	34.9%	20.8%
Last year	-2.4%	7.5%	4.7%

Source: Taxi Statistics, Department for Transport

Wales

Table 24: Taxis, PHVs and Drivers since 2004/05 – Wales

Year to 31 March	Taxis	PHVs	Total Licensed Drivers
2005	3.9	4.2	11.4
2007	4.7	3.7	11.3
2009	4.9	4.6	12.2
2011	5.0	4.1	12.3
2013	4.9	4.0	11.7
2015	5.1	4.2	11.7
2017	5.1	4.8	11.5
2018	5.0	4.9	12.3
2019	5.0	5.4	12.4
Since 2004/05	6.6%	44.9%	9.6%
Since 2013	0.4%	35.0%	5.6%
Last year	-1.0%	10.0%	0.6%

Source: Taxi Statistics, Department for Transport

In Wales, the taxi market has remained much more stable, the number of vehicles growing by 6.6% since 2004/05, but with a small fall in 2019. The PHV market has seen licensed vehicles increase by 45% over the same period, with a further increase of ten per cent in 2018/19. The total number of drivers has risen by just under ten per cent.

Scotland

In Scotland, meanwhile, the number of taxis increased by some 61% between 1985/86 and 2017/18, the last for which figures were available at the time of writing. Growth since 2005 has been very small (3.1%), with a 3% reduction in the total since the 2011 peak. The number of licensed taxi drivers is 13% lower than the peak seen in 2010.

In the PHV market, there has been growth, but much less spectacular than in England: in 2018, the number of licensed vehicles was 38% higher than in 2004/05, with a 32.6% increase in the number of drivers.

Table 25: Taxis, PHVs and Drivers since 1985 - Scotland

Year to 31 March	Taxis	Taxi Drivers	PHVs	PHV Drivers
1986	6.4	14.4		
2000	7.7	20.2		
2001	9.3	24.8	9.1	
2002	9.3	24.8	9.1	
2004	9.6	21.6	9.4	12.8
2005	10.0	25.0	10.1	11.3
2007	10.4	24.9	10.6	11.5
2009	10.5	25.6	11.3	11.4
2011	10.7	25.5	10.7	11.3
2013	10.6	24.6	10.2	11.3
2015	10.5	20.5	10.7	10.5
2017	10.4	23.3	13.0	13.5
2018	10.4	22.6	13.8	14.9
% changes				
Period	61.8%	57.0%	-	-
Since 2004/05	3.1%	-9.7%	6.2%	-6.3%
Since 2013	-2.3%	-8.1%	35.3%	31.5%
Last year	-0.4%	-3.0%	4.4%	-7.1%

Source: PTIS analysis of Scottish Transport Statistics

Chapter 7: Why People Use Buses

7.1 Introduction

As we noted earlier, people do not travel for the sake of it. They need a reason to make their journey, and changes in these reasons over time will have a significant effect on the volume and distribution of demand (in terms of both destination and time of travel).

Trends in the use of bus services for different purposes are tracked by the DfT's annual National Travel Survey (NTS) and provide a useful picture of why people are travelling and how importance of these different purposes varies over time.

There are seven journey purposes used in the NTS questionnaires, and in the analysis that follows, we have consolidated these into six. These are:

- Commuting

- Business

- Education (including as escort)

- Shopping

- Other Escort

- Personal business

- Leisure trips

At a time of rapid social and economic change, shifts in these patterns can be significant, and have the capacity to transform the market in which bus companies operate. There is very little opportunity for operators to influence these changes – as the case study included below illustrates.

7.2 The Cinema: A Case Study of Social Change

An outstanding example of how social change has affected the history of the bus industry is provided by the question of cinema attendances.

Statistics collected by the Cinema Advertising Association[3] show that when records began in 1935, there were 912.3m visits, growing to a peak of 1,635 million in 1946 – an average of just over 33 visits per person per year (more than one a fortnight).

For many people, the only means of accessing the cinema was to go by bus or tram, and given the modal split of journeys in those days, it is likely that the bumper cinema attendances of the late 1940s and early 1950s generated at least a billion passenger journeys a year.

After that peak year of 1946, a long decline set in. By 1956, visits had fallen to just over one billion. Thereafter, the fall accelerated sharply: living standards improved and the ownership of televisions grew sharply following the 1953 Coronation and the launch of ITV in 1955. Thus, cinema attendance halved between 1956 and 1960, halved again by

---

[3] *See www.cinemauk.org.uk/facts-and-figures/admissions/annual-uk-cinema-admissions-1935-2016.*

1968 and again by 1975. The nadir was reached in 1984 when the total attendance was just 54 million – 97% below the 1946 figure.

The result was the loss of a huge number of evening bus trips. This significantly damaged the performance of bus companies, who lost huge amounts of off-peak patronage and revenue. However, they had little or no opportunity to influence the changes.

After 1984, attendances recovered, and reached a peak of 173.5 million in 2009, just before the recession. Since then attendances have run at or slightly below 170 million a year, hitting 177 million in 2018. Meanwhile, the growth of multiplex cinemas, which began in Milton Keynes in 1985, has widened choice and appeal. By 2000, there were 225 multiplexes in the UK, boasting 1,830 screens. By 2017, this had grown further, to 395 sites (48% of the UK total) with 3,524 screens.[4]

The changes have fuelled a move away from town and city centres to out of town or edge of centre sites, much more difficult to serve successfully by public transport. Even so, by 2014, 48% of screens remain in town and city centres – a proportion which had risen again from 38% a decade earlier. Out of town sites accounted for 34% and edge of centre a further 14%.

7.3   Historic Trends in Trip Rates

The trends since the turn of the century are indicated by the number of trips per person for each journey purpose, as recorded in the different surveys since 2002. The analysis shows some significant changes over the period since the turn of the century.

The first and most glaring is the substantial fall in the use of buses for all journey purposes, with a fall of over 29% since the 2002 survey, and of 21% over the last five years. The largest fall in trip rate has been for shopping, now a hefty 42% down. Over the last five years, the rate has fallen by over 27%.

This has been mirrored by a fall in personal business trip rates of some 23% since 2002 – though there had previously been some growth in this sector, so that the fall is much higher over the last five years, at almost 33%. This reflects such things as the movement of banking and other financial services online and the closure of bank and building society branches.

Whereas there had previously been some stability in some segments – notably commuting, business trips and education, these too have shown sharp declines over the last five years – though the trip rate for business remains higher than in 2002.

---

[4] See www.cinemauk.org.uk/facts-and-figures

Table 26: Use of Local Bus Services outside London by Journey Purpose
Trips per person per year

Survey	Commuting	Business	Education	Shopping	Other Escort	Personal Business	Leisure	All Purposes
2002	8.8	0.3	8.1	14.4	1.6	4.2	8.8	46.2
2003	9.4	0.5	8.9	13.5	1.5	4.4	8.7	47.0
2004	9.3	0.5	7.5	14.5	1.2	3.8	8.2	45.1
2005	8.6	0.4	7.7	12.3	1.2	4.3	8.3	42.8
2006	9.0	0.3	8.2	14.2	1.2	4.2	9.1	46.3
2007	8.0	0.7	7.5	13.4	1.1	4.1	9.1	43.9
2008	7.9	0.6	7.2	14.1	1.3	4.5	8.5	44.2
2009	7.7	0.5	7.1	14.1	1.6	4.8	9.3	45.2
2010	7.1	0.5	7.7	13.3	1.2	4.2	8.2	42.2
2011	7.1	0.5	7.7	12.2	1.5	4.3	8.7	42.0
2012	7.4	0.6	8.3	11.3	1.4	4.1	8.2	41.3
2013	7.0	0.6	8.0	11.4	1.2	4.7	8.5	41.5
2014	7.5	0.6	7.5	11.4	1.4	4.2	7.9	40.5
2015	7.7	0.6	7.8	11.3	1.5	4.1	8.4	41.3
2016	6.5	0.4	7.0	9.7	1.0	3.1	7.7	35.4
2017	7.3	0.6	7.1	9.7	1.2	3.7	7.8	37.4
2018	5.9	0.4	6.7	8.3	1.2	3.2	7.0	32.7
% change								
Since 2002	-33.3%	43.1%	-17.4%	-42.2%	-26.8%	-23.7%	-20.7%	-29.3%
Last Five	-16.2%	-26.7%	-16.6%	-27.1%	-5.4%	-32.8%	-17.8%	-21.3%
Last Year	-9.3%	2.0%	-4.0%	-14.4%	17.1%	2.7%	-9.4%	-7.8%

Source: Department for Transport National Travel Surveys for the year shown, analysed by PTIS.

7.4 Historic Trends in Passenger Journeys

The NTS data enables us to take a snapshot of the relative importance of each journey purpose to the bus industry.

We can estimate this by applying these trip rates to the wider population to give us the percentage breakdown for each year, using that year's NTS results and the Mid-Year Population Estimates. These percentages can then be applied to the number of trips recorded in each year to give us an overall view of bus patronage by journey purpose.

The relative importance of each purpose is shown Table 28 below. This clearly shows the diminishing importance of shopping as a reason for using the bus, with the proportion falling from 38.5% in 2002 to 34% in 2012 and then to 32.4% in 2019. On the other hand, this has been offset trips for educational purposes have become steadily more important. Leisure trips, too, have become more important, accounting for 27% of trips in 2018, up from 23.5% in 2002. The graph at Figure 7-1 shows the changes.

Table 27: Bus Journeys in England outside London by Journey Purpose (%)

Trips	Commuting	Business	Education	Shopping	Other Escort	Personal Business	Leisure	Total
2002	23.6	0.8	21.7	38.5	4.2	11.2	23.5	100.0
2003	24.7	1.4	23.1	35.4	4.0	11.4	22.8	100.0
2004	25.4	1.5	20.2	39.4	3.2	10.3	22.4	100.0
2005	24.8	1.2	22.4	35.5	3.5	12.5	24.0	100.0
2006	24.3	0.9	22.1	38.3	3.2	11.3	24.6	100.0
2007	23.0	2.0	21.4	38.7	3.1	11.7	26.3	100.0
2008	22.3	1.8	20.2	39.5	3.7	12.5	24.0	100.0
2009	21.4	1.5	19.7	39.4	4.5	13.5	26.0	100.0
2010	20.9	1.3	22.8	39.2	3.4	12.4	24.1	100.0
2011	21.4	1.6	23.2	36.6	4.4	12.9	26.2	100.0
2012	22.3	1.7	25.2	34.0	4.3	12.5	24.7	100.0
2013	21.3	1.7	24.3	34.6	3.7	14.3	25.7	100.0
2014	23.0	1.7	23.0	35.2	4.2	12.8	24.4	100.0
2015	23.3	1.8	23.6	34.3	4.4	12.5	25.7	100.0
2016	23.5	1.5	25.2	35.1	3.5	11.2	27.8	100.0
2017	24.7	1.9	24.2	32.8	3.9	12.5	26.5	100.0
2018	22.9	1.6	26.1	32.4	4.5	12.4	27.2	100.0

PTIS Estimates based on NTS Data and ONS Mid-Year Population Estimates and DfT Annual Bus Statistics

Figure 7-1: Bus Journeys (%) by Journey Purpose, England outside London
Snapshots in 2002, 2007, 2012 and 2018

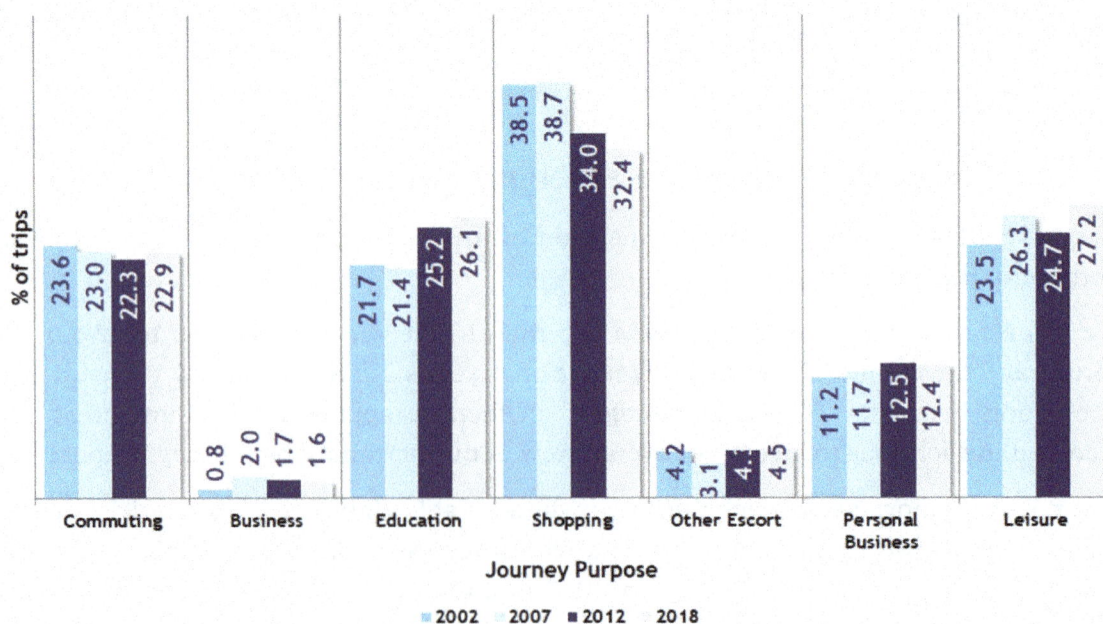

Table 28: Bus Trips in England outside London by Journey Purpose
Passenger Journeys (million)

Year to 31 March	Commuting	Business	Education	Shopping	Other Escort	Personal Business	Leisure	Total
2002	581	19	534	948	103	274	579	2,459
2003	602	33	564	862	98	278	556	2,437
2004	607	35	485	945	75	247	536	2,395
2005	581	28	524	831	83	291	560	2,338
2006	542	19	491	851	71	250	547	2,224
2007	514	44	479	864	70	261	587	2,233
2008	513	41	466	911	86	288	552	2,305
2009	508	36	466	934	107	319	615	2,370
2010	502	32	549	943	82	299	579	2,408
2011	509	38	551	868	104	305	622	2,375
2012	524	40	592	798	101	294	579	2,349
2013	493	39	564	802	85	332	596	2,316
2014	519	39	519	794	95	289	550	2,256
2015	534	42	540	785	102	287	587	2,289
2016	531	33	571	795	80	253	630	2,264
2017	548	42	536	726	87	277	587	2,215
2018	504	35	574	713	99	272	598	2,198
% change								
Since 2003	-13.2%	86.3%	7.6%	-24.7%	-4.7%	-0.7%	3.3%	-10.6%
Last Five	-1.7%	-14.2%	23.3%	-21.7%	14.6%	-5.4%	8.2%	-4.6%
Last Year	-2.9%	-10.5%	10.6%	-10.2%	4.1%	-5.7%	8.6%	-2.6%

PTIS Estimates based on NTS Data and ONS Mid-Year Population Estimates and DfT Annual Bus Statistics

7.5   Changes in Shopping Destinations

Even before the internet, the UK's retail industry has been in a continuous process of radical change. Since the mid-1980s, these changes in retailing practices have had a direct impact on the bus industry.

In particular, the growth of suburban retail centres and supermarkets on the outskirts of towns has reduced the number of people who travel by bus for shopping journeys. This was previously a significant market for the industry.

As a result, in 2006 the Commission for Integrated Transport (CfIT) published a report looking at the impact of the retail sector on delivering sustainable transport. This showed the proportion of shopping trips by mode and type of retail outlet. The results are summarised in Table 29.

It will be seen that any movement away from traditional town centres has a disproportionate effect on the number of people who use public transport, especially the bus. Although not shown in the table, a there is a similar trend for other leisure activities.

Table 29: Public Transport share of shopping trips by centre type

Destination	Car	Park & Ride	Public Transport	Walk & Cycle	Taxi & Other
Nearest Town / City Centre	44%	2%	30%	23%	2%
Other Town / City Centre	66%	2%	20%	12%	1%
Local Centre	49%	0%	10%	40%	1%
Out of Town Retail	85%	0%	7%	6%	1%
Edge of Town Retail	82%	0%	8%	9%	0%
Local Convenience Store	42%	0%	2%	54%	1%
Large Supermarket	81%	0%	5%	12%	1%

Source: Sustainable Choices and the Retail Sector, Commission for Integrated Transport 2006.

7.6 Journeys to Work

Information from the Labour Force Survey undertaken by the Office for National Statistics gives a picture of the proportion of the workforce which travels to work by bus. The question is asked every autumn.

Most Recent Data

The most recent data covers the October-December 2018 period – when the workforce stood at 31.8 million. According to this, 6.7% of the workforce across Great Britain used the bus to travel to work – a total of 2.14 million people. We would therefore estimate that the total volume of trips from journeys to and from work is therefore of the order of 1.03 billion.

The highest proportion was in London (14.0% overall, rising to 16.8% when looking at inner London). Only two other areas saw more than 10% of the workforce using the bus – in the metropolitan (PTE) areas of Tyne & Wear and West Midlands.

Less densely populated areas outside the big cities see a much lower use of the bus for work, with areas such as the North East outside Tyne & Wear and the parts of the North West outside the two big conurbations, where the proportions are as low as 2.9% and 3.7% respectively. The lowest proportion of all is that part of the West Midlands region which falls outside the Combined Authority area, at 1.6%.

Table 30 below shows the sector figures and Table 31 the full regional analysis.

Table 30: Commuting by Bus in Great Britain 2018, by Sector

Region of residence	% of workforce	Workforce (000s)	Users (000s)	Estimated Trips (000s)
London	14.0%	4,676	655	314,601
PTE Areas	9.2%	4,935	452	216,960
English Shires	3.8%	17,312	661	317,368
Scotland	9.3%	2,677	249	119,637

Wales	4.6%	1,529	71	33,849
TOTAL GB	6.7%	31,817	2,140	1,027,388
GB Outside London	5.5%	27,141	1,485	712,787

Source: PTIS analysis of Labour Force Survey 2018, Office for National Statistics, as published in Transport Statistics Great Britain 2019, Sheet TSGB0108. Trips estimated by 2 x single trips per day for 5 days a week, 48 weeks a year.

Table 31: Commuting by Bus in Great Britain 2018, by Region

Region of residence	% of workforce	Workforce (000s)	Users (000s)	Trips (000s)
North East	6.2%	1,221	76	36,359
Tyne and Wear	10.4%	527	55	26,386
Rest of North East	2.9%	694	20	9,594
North West	6.6%	3,498	231	111,083
Greater Manchester	9.1%	1,350	123	59,077
Merseyside	8.1%	688	55	26,607
Rest of North West	3.7%	1,459	54	25,761
Yorkshire and The Humber	7.1%	2,596	183	88,009
South Yorkshire	7.6%	647	49	23,656
West Yorkshire	8.5%	1,117	95	45,601
Rest of Yorkshire and The Humber	4.7%	832	39	18,668
East Midlands	4.5%	2,316	104	49,849
West Midlands	5.6%	2,754	154	73,786
West Midlands (Combined Authority)	10.0%	1,293	130	62,241
Rest of West Midlands	1.6%	1,461	24	11,301
East of England	2.8%	3,110	87	41,613
London	14.0%	4,676	655	314,601
Inner London	16.8%	1,966	331	158,896
Outer London	12.0%	2,710	324	155,637
South East	4.5%	4,624	207	99,366
South West	4.5%	2,816	128	61,215
England	6.6%	27,612	1,817	872,311
Wales	4.6%	1,529	71	33,849
Scotland	9.3%	2,677	249	119,637
Great Britain	6.7%	31,817	2,140	1,027,388

Source: PTIS analysis of Labour Force Survey 2018, Office for National Statistics, as published in Transport Statistics Great Britain 2019, Sheet TSGB0108. Trips estimated by 2 x single trips per day for 5 days a week, 48 weeks a year.

Historic Data

Looking backwards, we see that the proportion of people using bus to get to and from work has tended to fall over time. In Great Britain as a whole, it has fallen from 7.7% of the workforce in 2005 to 6.7% in 2018. However, this has largely been offset by a 12.9% increase in the workforce, rising from 28.2m to 31.8m. As a result, the volume of bus commuters and the trips they make have only fallen by 1.4%.

Outside London, the figures show a higher fall: bus use has fallen from 6.4% of the workforce to 5.5%, whilst the rise in the workforce has been lower, at 10.3%. Consequently, the number of users and volume of trips has fallen by 6.2%.

The details are shown in Table 32 and Table 33 below. The trends in estimated trips are shown in the graph at Figure 7-2. The figures all indicate a rather more stable commuting market than might have been expected from headline patronage figures. The dip caused by the 2008/09 financial crisis and recession can clearly be seen – as can the subsequent recovery until 2015. However, since then the decline has been more marked, especially in the non-London markets.

Table 32: Commuting by Bus in Great Britain 2005-18

Year	% of workforce	Workforce (000s)	Users (000s)	Trips (000s)
2005	7.7%	28,177	2,172	1,042,442
2006	7.8%	28,496	2,234	1,072,531
2007	7.5%	28,852	2,166	1,039,575
2008	7.5%	28,813	2,165	1,039,437
2009	6.9%	28,394	1,972	946,716
2010	7.0%	28,612	1,998	959,064
2011	7.5%	28,604	2,147	1,030,573
2012	7.3%	29,178	2,142	1,028,066
2013	7.2%	29,550	2,141	1,027,482
2014	7.3%	30,154	2,203	1,057,338
2015	7.2%	30,792	2,212	1,061,672
2016	7.1%	31,061	2,197	1,054,366
2017	6.8%	31,399	2,122	1,018,527
2018	6.7%	31,817	2,140	1,027,388
% change		12.9%	-1.4%	-1.4%

Table 33: Commuting by Bus in Great Britain outside London 2005-18

Year	% of workforce	Workforce (000s)	Users (000s)	Trips (000s)
2005	6.4%	24,600	1,584	760,140
2006	6.6%	24,848	1,651	792,515
2007	6.5%	25,120	1,623	778,885
2008	6.2%	24,973	1,551	744,321
2009	5.7%	24,615	1,402	673,168
2010	5.6%	24,737	1,397	670,356
2011	6.3%	24,740	1,568	752,708
2012	6.0%	25,097	1,497	718,701
2013	6.0%	25,378	1,519	729,287
2014	6.0%	25,866	1,542	740,331
2015	5.9%	26,359	1,560	748,635

2016	5.7%	26,485	1,514	726,636
2017	5.6%	26,716	1,502	721,097
2018	5.5%	27,141	1,485	712,787
% change		10.3%	-6.2%	-6.2%

Source for Tables 30 and 31: PTIS analysis of Labour Force Survey 2018, Office for National Statistics, as published in Transport Statistics Great Britain (Time Series in sheet TSGB0108). Trips estimated by 2 x single trips per day for 5 days a week, 48 weeks a year.

Figure 7-2: Commuting by Bus in Great Britain 2005-18

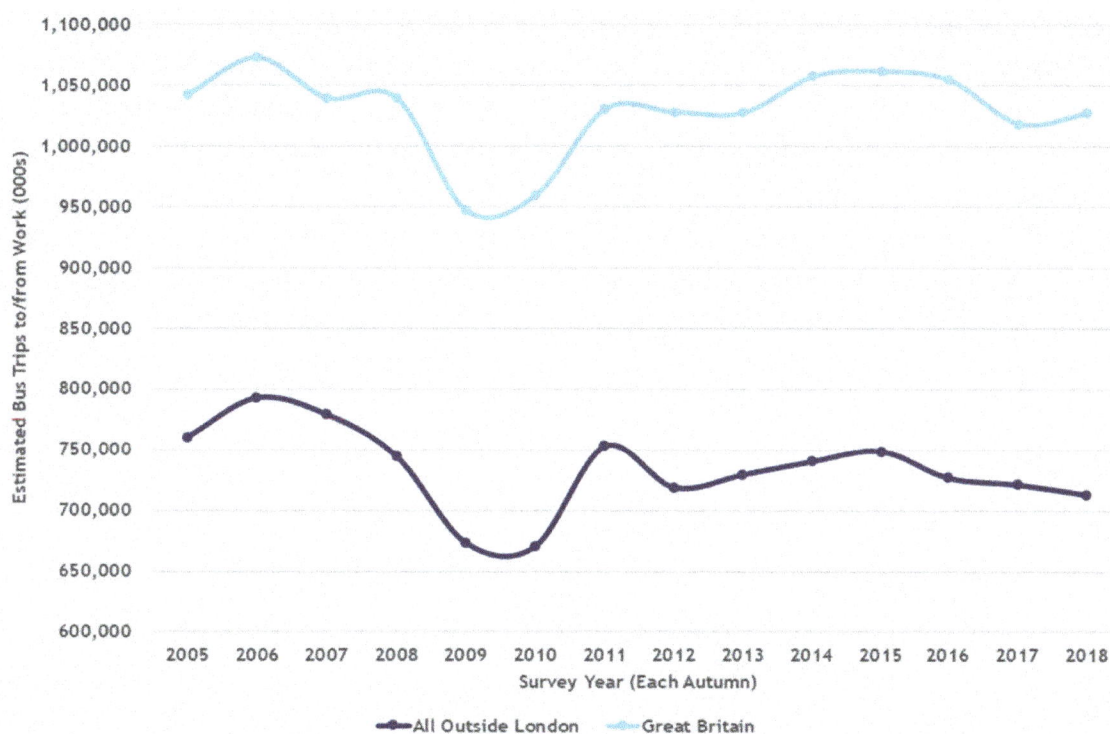

Source: PTIS analysis of Labour Force Survey 2018, Office for National Statistics, as published in Transport Statistics Great Britain 2019, Sheet TSGB0108. Trips estimated by 2 x single trips per day for 5 days a week, 48 weeks a year.

7.7   Working from Home

A reduction in journeys to work has long been foreseen from the increase in home working made increasingly possible by the growth of reliable and fast internet connections.

New data on this was released in March 2020 by the Office for National Statistics, showing a clear trend towards more home working.

According to the Annual Population Survey for 2019, 14.5 per cent of the workforce reported that they worked from home, from an office in the same grounds or building, or used home as a base. This compared with 14.4% in 2015. This represents a total of around 4.7 million people out of a total workforce of 32.6 million.

In the most recent survey, respondents were asked whether they had worked from home at some time during the last week. Four million said they had. A further question also asked respondents if they ever worked from home. This time, ONS received a positive response from 8.7 million. Unfortunately, there is no time series data for this.

The ONS analysis also goes on to analyse the industries and the occupations which feature most amongst home workers, and inevitably the jobs are the more senior ones – managers, directors and senior officials (category 1) of whom almost half (46.7 per cent) sometimes work from home, a slightly lower proportion in professional occupations (category 2 – 45 per cent) and still a substantial portion of those in associate professional and technical occupations (category 3 – 36.5 per cent). In administrative and technical occupations in category 4 the proportion is still quite high at 19.9 per cent.

As we saw in Table 17 above, a significant proportion of bus trips continue to be made by people in these occupations – and they are precisely the groups amongst which there will have been a significant expansion of home-working during the COVID-19 lockdown.

The expectation is that the enforced switch to working from home will lead to significant increases in the proportion of the workforce who do this on a more frequent or even permanent basis. The press has already carried stories of firms laying plans to either close, or significantly reduce the size of, their head offices – relying on video conferencing, e-mailing and cloud file sharing systems to provide the bulk of their communications.

Early research by consultants LEK[5] carried out at the beginning of April 2020 suggested that, after the lockdown, 46% of respondents wanted to work from home more than before, including 17% who wanted to work entirely from home in future (a figure which rose to 21% for those under 34).

---

[5] *https://www.linkedin.com/pulse/how-commuters-react-home-working-experience-survey-data-andrew-allum/?tracki*

Chapter 8: The PTE Areas

8.1 Introduction

The original Passenger Transport Executives were established by the passage of the 1968 Transport Act. This established Passenger Transport Authorities and Passenger Transport Executives in the main UK conurbations of West Midlands, Greater Manchester, Tyneside and Merseyside.

In each case, the PTE became the operating bus company in the territory, absorbing the existing municipal undertakings with their areas. Other operations were then acquired by agreement.

On local government reorganisation in 1974, the new metropolitan county councils became the Passenger Transport Authorities for their area. Two further PTEs were established, in South Yorkshire and West Yorkshire, while the existing PTEs expanded by taking over other local authority transport companies within their expanded areas.

Following the abolition of the Metropolitan County Councils in 1987, the PTAs once more became independent organisations, with membership nominated by the constituent Metropolitan District Councils.

Following the 2008 Local Transport Act, the Passenger Transport Authorities were renamed as Integrated Transport Authorities. However, this proved to be short-lived, as the Act was overtaken during the following year by the Local Democracy, Economic Development and Construction Act 2009. This provided for the establishment of Combined Authorities.

The new bodies were designed to be created, on a voluntary basis, in areas where they were considered likely to improve transport, economic development and regeneration. The legislation allows a group of local authorities to pool appropriate responsibility and receive certain delegated functions from central government in order to deliver transport and economic policy more effectively over a wider area.

So far, at the time of writing, seven Combined Authorities have been created, covering the six former PTE areas, with a seventh in the Tees Valley (covering Darlington, Hartlepool, Middlesbrough, Redcar & Cleveland and Stockton-on-Tees).

The North East authority was initially expanded from the core area of Tyne & Wear to cover Northumberland and County Durham. Plans were announced in 2015 to proceed with a devolution settlement and the election of a new Mayor for the combined authority in 2017. However, this deal collapsed in 2016 when Durham, Gateshead, South Tyneside, and Sunderland Councils withdrew their support. In 2018, new arrangements were agreed under which Newcastle City Council, North Tyneside Borough Council, and Northumberland County Council left the North East Combined Authority and formed the new North of Tyne Combined Authority. For transport purposes, the two authorities come together as the North East Joint Transport Committee.

Combined Authorities also now exist in Cambridgeshire and Peterborough and in the West of England (effectively recreating the old Avon County).

The six former PTE areas remain amongst the largest settlements in England outside London and are often seen as a unit. In fact, however, they all started with radically different demographic and economic characteristics, and their economic and social history since 1968 has varied widely.

8.2 Overview

In the years immediately after bus deregulation, the PTE areas saw particularly large falls in patronage. As well as the immediate fall associated with the network upheavals in 1985/86 and 1986/87, demand subsequently continued to fall sharply, the trends worsened by the onset of the early 1990s recession. The number of passenger journeys by bus in the PTE areas fell by 37.7% between 1987/88, and 2004/05, but over 26% of the that fall had already occurred by 1994/95.

Once the economic recovery began in the mid-1990s, demand began to stabilise; since then, the picture has been mixed, with some showing limited growth but others continuing falls.

There were several reasons for this early and steep fall:

- Before deregulation, these areas enjoyed the highest level of fares subsidy, resulting in particularly steep rises to achieve commercial rates. This varied between South Yorkshire, which had run a low fare policy, and West Midlands, with a comparatively low level of subsidy.

- Economic recession and the loss of population hit many of these areas hard over period, reducing overall demand for transport by all modes

- Car ownership rates in these areas have grown rapidly. Since the rates were lower than average at the start of the period, this expansion of car ownership has tended to have a disproportionate effect on patronage.

- The former PTA bus operations faced difficult hurdles after deregulation. Cost levels were substantially higher than ex-NBC operators and they had to adjust to the phasing out of political control, restructuring and downsizing. These companies dominated the market within the PTE areas, with over 80% of turnover altogether.

- There was steady erosion of the once generous concession fare schemes in several areas, which affected bus patronage.

Since 2004/05, demand across the PTE areas has both grown and then shrunk again. Growth took place in the aftermath of the introduction of the English National Concessionary Travel Scheme but was brought to a halt by the onset of the recession. Since then, demand has fallen quite sharply, so that the total in 2018/19 was 13.3% lower than in 2004/05, and 6.9% down in the last five years.

The number of passenger kilometres travelled by bus in the PTE areas has fallen rather more slowly that passenger journeys, suggesting some lengthening of average journeys.

The figures are shown in Table 34. Particularly noteworthy are the shifts in real-term revenue over the period. This peaked in 2012 and has been falling ever since. It is also noticeable that – thanks to cuts in both supported and commercial services - service supply has fallen rather more quickly than demand.

The trends are illustrated in graph form, using an index for each measure set at 2004/05=100 at Figure 8-1 below.

Table 34: Bus Market Statistics for the PTE Areas since 2005

Year to 31 March	Passenger Journeys (million)	Passenger Kilometres (Millions)	Passenger Revenue £m (2018/19 Prices)	Kilometres Run
2005	1,047	5,792	1,153	592
2006	1,049	5,707	1,138	588
2007	1,052	5,822	1,171	591
2008	1,067	6,166	1,165	597
2009	1,074	6,323	1,187	589
2010	1,062	6,251	1,203	569
2011	1,032	5,927	1,199	567
2012	1,004	5,704	1,205	563
2013	977	5,600	1,195	553
2014	991	5,593	1,179	546
2015	976	5,540	1,155	531
2016	949	5,257	1,135	518
2017	938	5,301	1,056	497
2018	919	5,080	1,100	483
2019	908	5,178	1,102	491
% changes				
Since 2005	-13.3%	-10.6%	-4.4%	-16.9%
Last Five Years	-6.9%	-6.5%	-4.6%	-7.4%
Last Year	-1.2%	1.9%	0.2%	1.8%

Source: Department for Transport Annual Bus Statistics

Figure 8-1: Index of Key Measures, PTE Areas

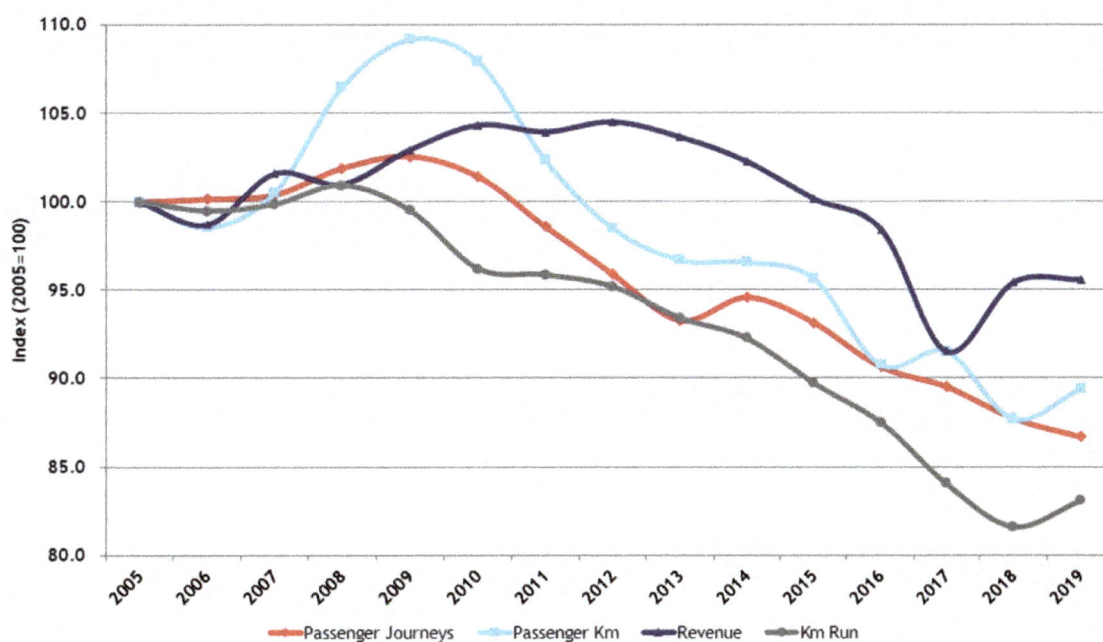

8.3   Commercial Performance

Analysis of this data produces some figures on yields and average loads which are shown in Table 35 below.

It will be seen that the estimates of passenger kilometres travelled, coupled with reductions in mileage operated, have resulted in a steady growth in the average bus load (calculated as passenger kilometres per vehicle kilometre), improving from around 9.8 at the start of the period to 10.5 in 2018/19. However, this remains well below the London levels, which are virtually twice these figures – reflecting the much greater population densities seen in the capital.

Though real-term revenue has fallen, it has done so at a slower rate than patronage, which means that fare increases have driven revenue per passenger journey, per passenger kilometre travelled and per vehicle kilometre operated have recovered from earlier falls. The adverse trends resulted from reduced concessionary fares reimbursement, changes in the market, including the spread of day and period tickets and the impact of the changes to travel patterns (for example by a shift from peak to off-peak travel at lower fares).

Table 35: PTE Bus Statistics - Market Analysis

Year to 31 March	Average Journey (Kilometres)	Average Fare (£, 2018/19 prices)	Yield (£, 2018/19 prices)	Average Load (Passenger Journeys per Passenger Km)	Revenue per kilometre run (£,2018/19 prices)
2005	5.53	1.101	0.199	9.79	1.95
2006	5.44	1.085	0.199	9.70	1.93
2007	5.54	1.114	0.201	9.86	1.98
2008	5.78	1.091	0.189	10.33	1.95
2009	5.89	1.105	0.188	10.74	2.02
2010	5.89	1.133	0.192	10.99	2.11
2011	5.74	1.161	0.202	10.45	2.11
2012	5.68	1.200	0.211	10.13	2.14
2013	5.73	1.223	0.213	10.13	2.16
2014	5.65	1.190	0.211	10.25	2.16
2015	5.68	1.184	0.208	10.44	2.18
2016	5.54	1.195	0.216	10.15	2.19
2017	5.65	1.126	0.199	10.66	2.12
2018	5.53	1.197	0.217	10.52	2.28
2019	5.70	1.214	0.213	10.54	2.24
% changes					
Since 2005	3.1%	10.3%	6.9%	7.6%	15.1%
Last Five Years	0.4%	2.6%	2.1%	1.0%	3.1%
Last Year	3.2%	1.4%	-1.7%	0.1%	-1.6%

Source: PTIS analysis of DfT Bus Statistics

8.4   Individual PTE Markets

The figures for the six individual areas are shown in Table 36 below. These are taken from the PTEs' own statistical returns which may not agree with those submitted by bus operators used in Table 34.

The fastest decline in patronage since 2004/05 has been in West Yorkshire, where 29.2% of passengers have been lost. This is followed by Merseyside (23.4%), South Yorkshire (19.5%), West Midlands (16.6%) and Greater Manchester (13.2%). The smallest fall has been in Tyne and Wear at 4.3%.

Over the most recent five-year period, West Yorkshire has seen the fastest fall, at 24.8%, followed South Yorkshire on 16.1%. Tyne and Wear has seen the smallest fall at 3.3%.

Table 36: Bus Passenger Journeys in Individual PTE Areas
Millions of Passenger Journeys, PTE returns

Year to 31 March	Tyne & Wear	West Yorkshire	South Yorkshire	Merseyside	Greater Manchester	West Midlands
2005	134	196	111	164	218	314
2006	129	195	113	163	216	309
2007	133	197	115	154	223	316
2008	134	193	115	140	224	324
2009	140	195	118	149	235	327
2010	143	184	115	143	226	320
2011	141	178	113	142	224	306
2012	139	184	111	137	219	284
2013	140	180	105	136	220	276
2014	135	182	109	137	217	279
2015	133	184	107	137	211	275
2016	128	171	103	135†	209	267
2017	129	151	101	147	202	261
2018	122	141	93	142†	194	253
2019	128	139†	90	126	189	262
% changes						
since 2004/05	-4.3%	-29.2%	-19.5%	-23.4%	-13.2%	-16.6%
Last five	-3.3%	-24.8%	-16.1%	-7.9%	-10.4%	-4.7%
Last year	5.4%	-2.1%	-3.6%	-11.1%	-2.7%	3.8%

Source: Annual Bus Statistics, Sheet B0109b. May not be compatible with total figures shown in Table 35.
† - missing returns. Figures estimated by reference to operator returns for the same year recorded in BUS0109a.

8.5   Benchmarking Demand

One of the best methods of comparing demand in the different PTE areas is to consider the number of bus trips per head of the population. These data have been prepared by us for some years, and since been adopted by the Department for Transport. As a result, that we can not only consider the current position but also recent trends.

Table 9 provides a snapshot of the figures for each of the English PTE areas. The graph at Figure 8-2 shows the position in 2015/16 and compares it with the peak level seen in 2008/09, achieved after the introduction of the England-wide free concessionary fares scheme.

Table 37: Bus Ridership per Head of Population, Metropolitan Areas
Trips per person per year

Year to 31 March	Tyne & Wear	West Yorkshire	South Yorkshire	Merseyside	Greater Manchester	West Midlands
1986	248	141	259	235	116	182
1992	195	115	136	132	101	148
1996	153	108	126	117	90	139
2001	137	93	106	103	83	139
2004	121	85	89	104	87	127
2005	124	93	87	120	86	122
2006	118	92	88	119	85	119
2007	123	91	89	114	87	122
2008	123	88	88	109	88	124
2009	128	88	91	110	91	125
2010	129	83	87	106	87	121
2011	126	79	85	105	85	115
2012	126	83	83	99	82	104
2013	126	80	78	98	81	100
2014	121	81	80	98	80	100
2015	119	81	78	98	77	98
2016	114	75	75	97	76	94
2017	115	66	73	105	72	91
2018	108	61	67	100	69	87
2019	113	60	64	88	67	90
% changes						
Since 1985/86	-54.5%	-57.6%	-75.4%	-62.3%	-42.3%	-50.6%
Since 1995/96	-26.1%	-44.6%	-49.3%	-24.3%	-25.2%	-35.3%
Since 2004/05	-8.6%	-35.7%	-26.7%	-26.4%	-21.7%	-26.3%
Since 2008/09	-11.7%	-32.4%	-29.6%	-19.7%	-26.2%	-27.9%
Last Year	4.7%	-2.6%	-4.3%	-11.5%	-3.2%	3.1%

Source: PTIS analysis of Annual Bus Statistics, DfT, Sheet Bus0109b and ONS Mid-Year Population Estimates

For most of the last 30 years, Tyne & Wear has enjoyed the highest bus ridership of the PTEs, with the figure being as high as 248 at the time of deregulation in 1986. It fell to a low of 118 in 2005/06, before recovering to 129 in the peak year of 2009/10. Since then, though, a combination of the recession and social change has driven numbers back downwards to 113, a fall of 12%.

The West Midlands had the next highest ridership at 90. At the time of deregulation, this area was only fourth highest, with a figure of 182. However, the area did not experience the sharp falls seen in the other areas during the 1990s, so that by the middle of the decade it was second highest, and for a period in the middle of the "noughties" vied for the highest ridership with Tyne & Wear. The authorities had maintained free off-peak concessionary travel for its residents throughout the period (one of the reasons for the relative stability of the market in the 1990s), so that the "bounce" from the national free scheme was much lower. In 2008/09, ridership stood at 125 journeys per person per year. Since then, it has fallen by 29.1% to stand at 90 in 2018/19.

Next comes Merseyside. Back in 1986, it enjoyed the third highest ridership level, at 235, but this had more than halved by the mid-1990s, and fell to a low point of 103 in 2000/01. The peak in 2008/09 was 110. Since then, however, it has fallen by almost 17.4%, and stands at 88.

At the time of deregulation, Greater Manchester already had the lowest ridership figure, at 116 – less than half that in neighbouring Merseyside, in South Yorkshire or Tyne & Wear. The rate therefore tended to fall less steeply during the 1990s, and the low point was reached in 2000/01 of 83. It then recovered slightly and peaked at 91 in 2008/09. Since then, though, it has fallen back by 26% and stood at 67 in 2018/19.

South Yorkshire was next. The councils' low fares policy in the 1980s meant that it entered the deregulated period with ridership of 259, but it quickly fell, and continued to do so throughout the 1990s, reaching a low point of 87 in 2004/05. By 2008/09, it had recovered to 91, but has since fallen back by 29.6% and stood at 64 in 2018/19.

West Yorkshire now has the lowest ridership per head. It entered the deregulated era with a figure of 141, and this has fallen virtually every year since, without a significant bounce in 2008/09. It then stood at 88 and has since fallen by a further 32.4% to 60 in 2018/19.

Figure 8-2: Bus Ridership per Capita, PTE Areas 2008/09 & 2018/19

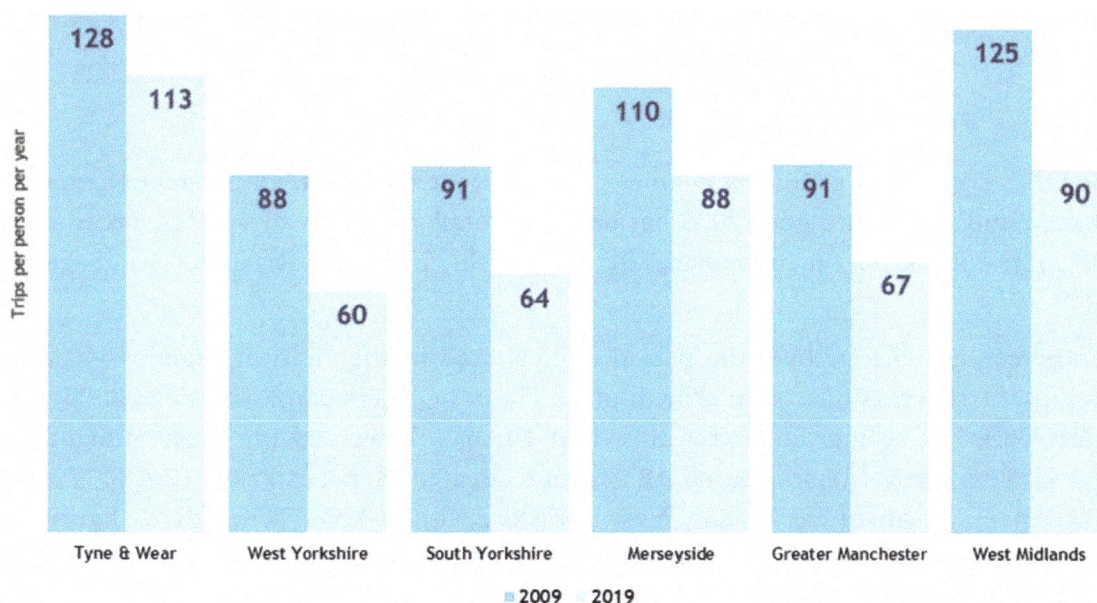

Figure 8-2: Bus Ridership per Capita, PTE Areas 2008/09 & 2018/19

8.6   Demographic and Economic Data

8.6.1   Population

In the immediate post deregulation period, population in almost all the PTE areas was falling. The most notable example of this was Merseyside, which experienced a fall of 11.3% in population between the 1981 census and the turning point in 2008, when the population has started to grow again. Other falls included Tyne & Wear (6.2%, 1981-2003), South Yorkshire (3.9%, 1981-2000), Greater Manchester (3.9%, 1981-2000), West Midlands (3.8%, 1981-2000). West Yorkshire did experience some decline between 1981 and 1985, and again between 1991 and 2000, but has otherwise seen growth in most years.

The chart below at Figure 8-3 illustrates the change in population between 2005 and 2018, using ONS Mid-Year Population Estimates. West Midlands has seen the fastest growth over that period, at 12.5%. Greater Manchester has experienced growth of 10.4% and West Yorkshire 9.5%. At the other end of the scale comes Merseyside on 4.3% and Tyne & Wear on 3.8%.

Figure 8-3: Population Change, PTE Areas since 2005

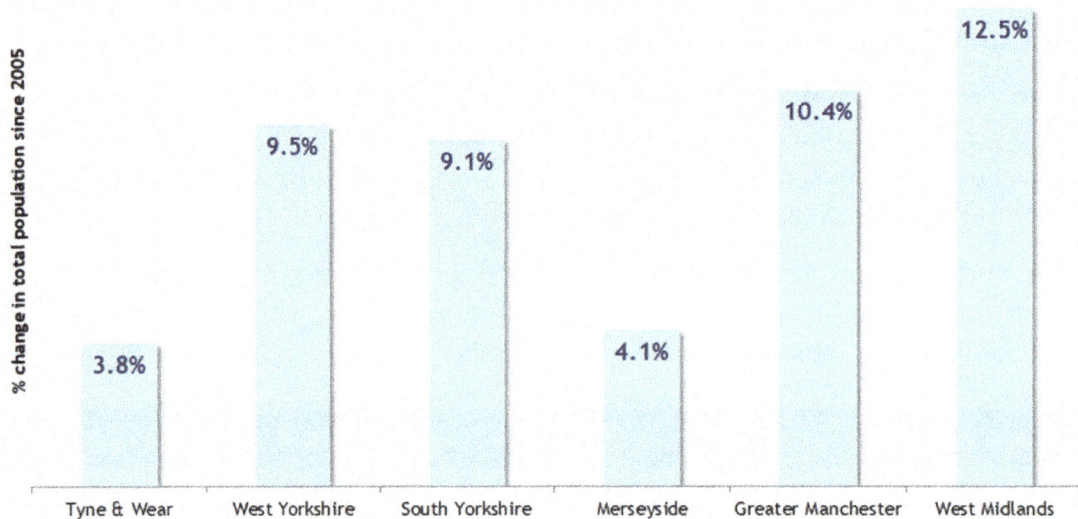

8.6.2   Employment

The ONS Labour Force Survey enables us to track the change in employment across the country, and the picture since 2005 has been of overall growth in all the PTE areas, albeit with some falls during the immediate aftermath of the financial crisis in 2008 and again in 2012.

The percentage change over the period is illustrated in the chart at Figure 8-4 below. Overall, the six areas have seen growth of 12.1% in numbers employed since 2005, with South Yorkshire seeing the largest growth at 14.6% followed by the West Midlands on 14.4%. Next come Merseyside on 12.9% and Greater Manchester on 12.4%. Lowest growth in employment has been in West Yorkshire, at just 8.3%. Though the figures are positive, they stand in stark comparison to London, where employment has risen by 32% over the same period, creating 1.1 million jobs in that time.

Figure 8-4: Population Economically Active & In Employment: % change

% change in Employment, 2005-2019

Authority	% change
Tyne & Wear	9.5%
West Yorkshire	8.3%
South Yorkshire	14.6%
Merseyside	12.9%
Greater Manchester	12.4%
West Midlands	14.4%
All PTE Areas	12.1%

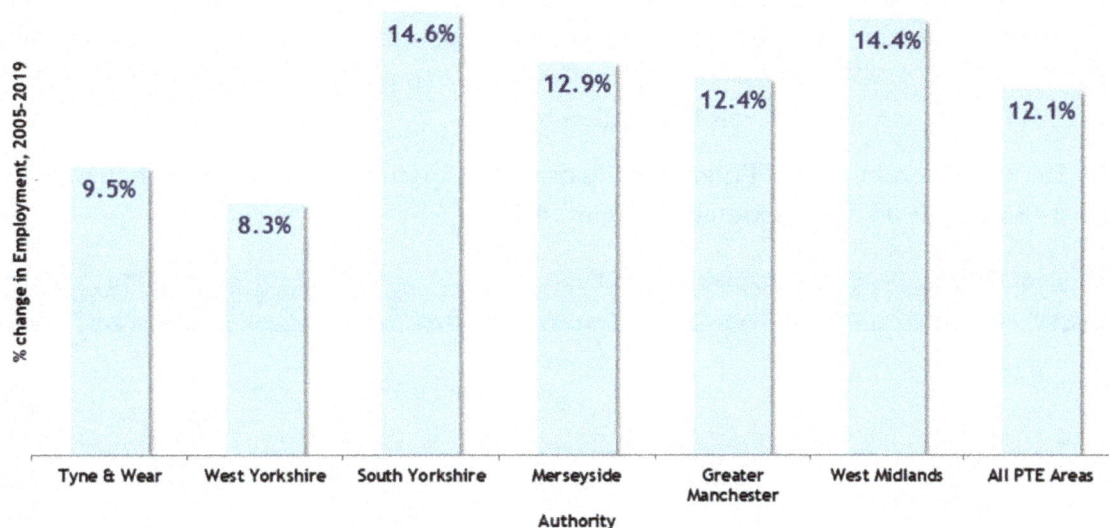

Source: NOMIS Database (ONS/University of Durham).

8.6.3   Economic Growth

The ONS figures on Gross Value Added (GVA) for each area shows a picture of the economic growth achieved in each area over the period between 2005 and 2018. The real-term change (adjusting for inflation using the GDP Deflator) over the whole period is measured for each area in Figure 8-5 below.

Figure 8-5: Gross Value Added in Real Terms, 2005-2018: % change

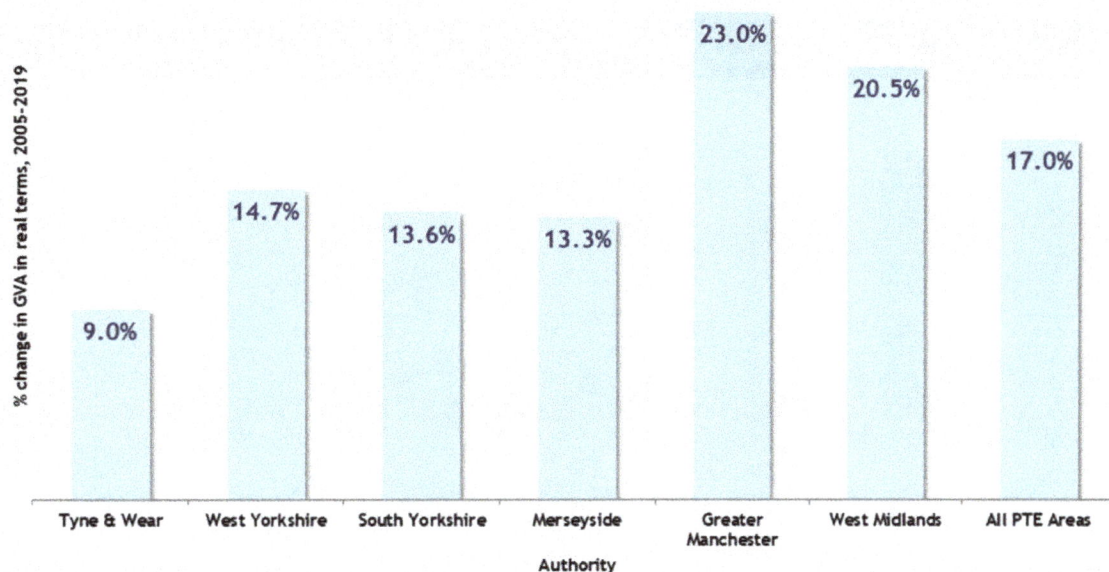

% change in GVA in real terms, 2005-2019

Authority	% change
Tyne & Wear	9.0%
West Yorkshire	14.7%
South Yorkshire	13.6%
Merseyside	13.3%
Greater Manchester	23.0%
West Midlands	20.5%
All PTE Areas	17.0%

Growth across all the areas has averaged 17% over the period, with Greater Manchester achieving the highest growth at 23%. This is followed by West Midlands on 20.5%, with the other four some way behind. Tyne & Wear saw the lowest real-term growth of 9.0%. As an illustration of the regional imbalances in the UK economy so often discussed, the equivalent figure for London was 42%.

8.6.4   Car Ownership

Car ownership grew substantially in all the PTE areas between 1986 and 2018. The highest level of growth has occurred in Tyne & Wear (77.5%), Merseyside (71.9%), with the lowest in Greater Manchester (51.1%). In the period since 2005, growth has been greatest in Tyne & Wear (8.3%), West Yorkshire (8.2%) and South Yorkshire (5.4%). In West Midlands, ownership levels have fallen by 0.2%.

The 2018 position in each PTE area is illustrated in Figure 8-6, whilst the changes since both 1986 and 2005 are illustrated in Figure 8-7.

Figure 8-6: Car Ownership per 1,000 Population in PTE Areas, 2018

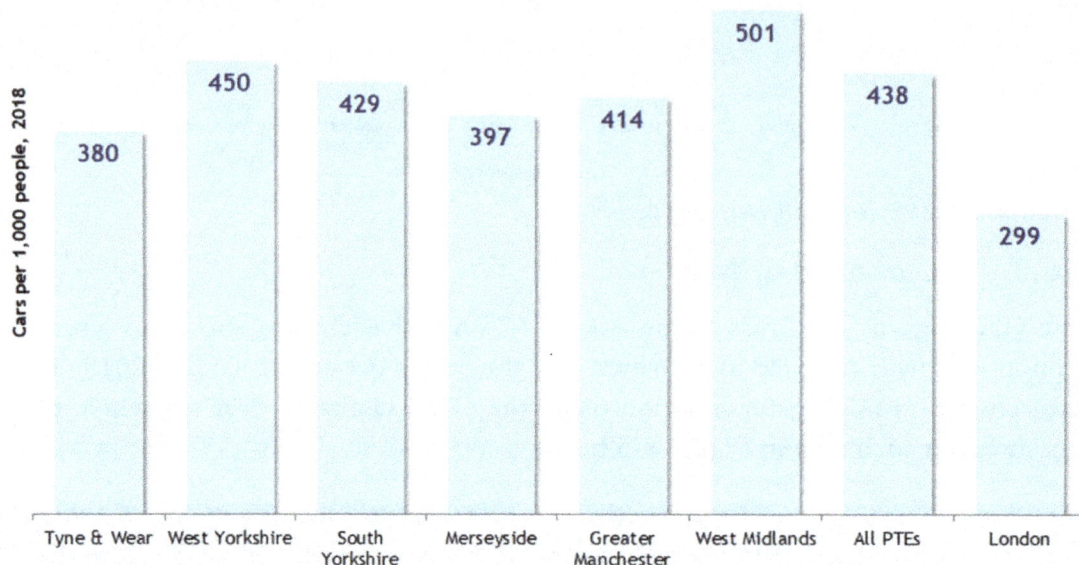

Figure 8-7: Change in Car Ownership, PTE Areas, since 1984 and 2005

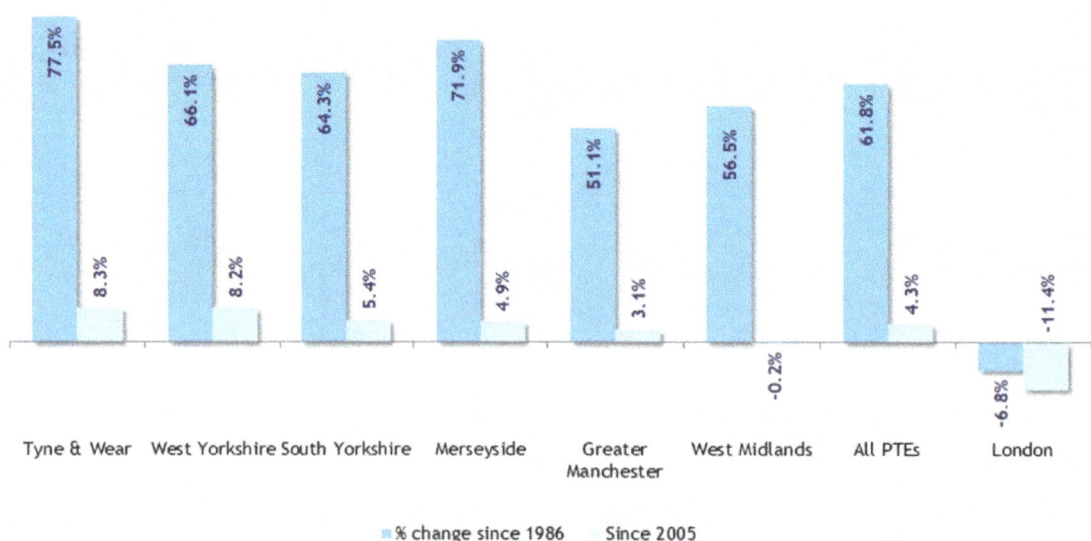

■ % change since 1986   Since 2005

PTE Area	Tyne & Wear	West Yorkshire	South Yorkshire	Merseyside	Greater Manchester	West Midlands	Tyne & Wear
1986	214	271	261	231	274	320	271
1994	267	326	338	287	349	407	341
2001	290	355	339	303	409	441	373
2005	347	406	399	370	423	495	420
2006	351	416	407	378	401	502	421
2007	354	408	405	382	401	478	414
2008	358	416	408	386	409	482	420
2009	360	421	411	389	407	481	421
2010	357	414	409	389	431	473	423
2011	352	410	407	386	422	481	420
2012	355	416	403	375	399	492	418
2013	356	416	402	373	396	483	415
2014	358	406	407	377	400	468	411
2015	363	411	414	382	406	476	418
2016	368	418	418	386	409	487	424
2017	373	431	423	390	405	500	430
2018	377	440	425	393	409	504	435
% changes							
Since 1986	77.5%	66.1%	64.3%	71.9%	51.1%	56.5%	61.8%
Since 1994	42.4%	38.2%	26.9%	38.0%	18.7%	23.1%	28.7%
Since 2001	31.0%	26.9%	26.4%	30.8%	1.3%	13.4%	17.6%
Since 2005	8.3%	8.2%	5.4%	4.9%	3.1%	-0.2%	4.3%

Table 38: Car Ownership per 1,000 Population, PTE areas (1986-2018)

8.7 Conclusions

As can be seen from the data in this chapter, the conditions for growth in bus patronage – in terms of growth in population, employment and the wider economy – have existed to some extent in these areas throughout the last fifteen years. Though car ownership rates have continued to grow, the percentage changes have been moderate, and the growth was interrupted by the hiatus following the 2008 financial crisis.

Data from the National Travel Survey suggests that, across the urban conurbations as a whole, the number of households not owning a car has remained remarkably consistent over the period since 2005, varying between 32% at the start of the period and 36% at the height of the recession. The 2018 figure was 34%.

Thus, we must look elsewhere for the reasons for the significant decline in patronage that has taken place across these areas. This is discussed further in Chapter 13 below.

Chapter 9: The English Shire Counties

9.1   Overall Trends

As was seen from Table 2 above, the overall picture in the shire areas of England means that, since the introduction of free concessionary travel, demand has recovered from the falls of the early years of the century to levels last seen in the mid-1990s. Despite the setbacks of the last couple of years, passenger numbers remain higher than before 2008.

The overall statistics for the shire market since 2004/05 are shown in Table 39 below.

Table 39: Bus Market Statistics for the Shire Areas since 2005

Year to 31 March	Passenger Journeys (million)	Passenger Kilometres (Millions)	Passenger Revenue £m (2018/19 Prices)	Kilometres Run
2005	1,177	7,914	1,398	1,061
2006	1,184	7,842	1,435	1,071
2007	1,253	8,407	1,567	1,067
2008	1,302	9,051	1,664	1,067
2009	1,334	9,496	1,783	1,077
2010	1,313	9,230	1,807	1,072
2011	1,317	9,082	1,820	1,073
2012	1,312	9,111	1,818	1,054
2013	1,279	9,503	1,836	1,048
2014	1,298	9,864	1,854	1,046
2015	1,288	9,664	1,869	1,040
2016	1,266	9,082	1,848	1,009
2017	1,260	8,968	1,826	1,005
2018	1,232	8,634	1,730	956
2019	1,213	9,119	1,696	932
% changes				
Since 2005	3.0%	15.2%	21.4%	-12.1%
Last Five Years	-5.9%	-5.6%	-9.2%	-10.3%
Last Year	-1.5%	5.6%	-1.9%	-2.5%

As can be seen, though patronage levels have been falling back since a peak reached in 2012, patronage and revenue remain above 2004/05 levels. Over the last five years, patronage has fallen by almost 6%, passenger km travelled by 5.6% and real-term revenue by 9.2%. Supply in terms of kilometres run has fallen by 10.3% over the same period. The trends are illustrated in the graph at Figure 9-1 below, which indexes each measure at 100 in 2004/05.

Figure 9-1: Market Trends for Shire Areas since 2005

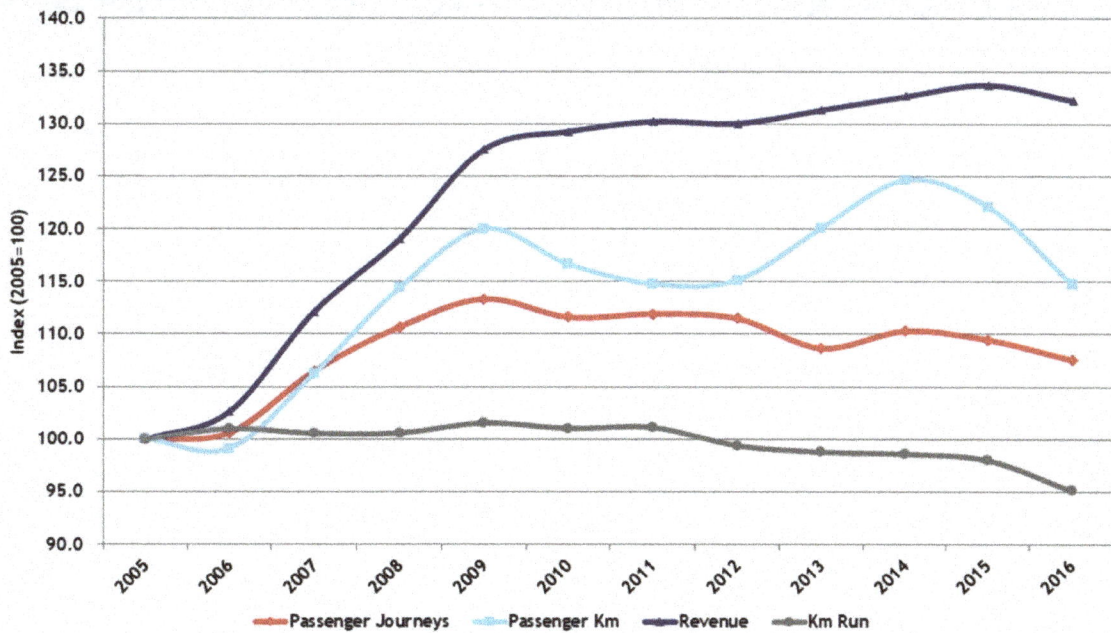

Figure 9-1: Market Trends for Shire Areas since 2005

9.2 Commercial Performance

The measures of performance are shown in Table 40 below. As might be expected given the rural or inter-urban nature of many shire areas, average journey lengths tend to be longer, and therefore average fares are higher, than in the big conurbations. Indeed, the introduction of free concessionary travel seems to have prompted, or at least coincided with, a significant increase in average journey length.

In contrast to the urban areas, earnings per passenger journey and per passenger kilometre have fallen back in recent years, with reductions of 3.6% and 3.8% respectively. The 2019 figure for average fare was below £1.40 for the first time since 2012.

A combination of longer journeys and reductions in service levels have driven the average load on each bus higher, having fallen to a low point of 7.3 in 2005/06. The most recent year showed a figure of 9.78.

Table 40: Shire Bus Statistics – Market Analysis

Year to 31 March	Average Journey (kilometres)	Average Fare (£, 2018/19 prices)	Yield (£, 2018/19 Prices	Average Load (Passenger Kms per Kilometre run)
2005	6.73	1.188	0.177	7.46
2006	6.62	1.212	0.183	7.32
2007	6.71	1.251	0.186	7.88
2008	6.95	1.278	0.184	8.48
2009	7.12	1.337	0.188	8.82
2010	7.03	1.376	0.196	8.61
2011	6.90	1.382	0.200	8.47
2012	6.95	1.386	0.200	8.64
2013	7.43	1.436	0.193	9.07
2014	7.60	1.428	0.188	9.43
2015	7.50	1.451	0.193	9.30
2016	7.18	1.460	0.203	9.00
2017	7.12	1.449	0.204	8.93
2018	7.01	1.404	0.200	9.03
2019	7.52	1.399	0.186	9.78
% changes				
Since 2005	11.8%	17.8%	5.3%	31.1%
Last Five Years	0.2%	-3.6%	-3.8%	5.2%
Last Year	7.3%	-0.4%	-7.1%	8.3%

9.3   Regional Trends

As with other broad geographical markets, though, there are marked differences between different parts of the country, and the figures published by the DfT for each of the English regions allow us to understand these variations.

The figures for the shire areas of each Government Office region are shown in Table 41 below. In regions where PTEs exist, the recorded patronage for the PTE area has been deducted from the regional totals to provide an estimate for the shire areas only within that region. These must be treated with some caution but do provide a useful means of comparing overall trends in each area.

It will be seen that, in the period since 2005, several areas in the midlands and the north have continued to lose patronage whilst generally those in the southern part have grown. The two notable exceptions are the North West, where patronage has grown by 11.6% since 2004/05, and Yorkshire and the Humber, which has seen an increase of over 70%. The South West has gained 27.7% more passenger journeys, and the South East 23%. East of England's growth has been less strong at 5.8%.

The North East region has seen the largest fall in patronage, 43.3%, followed by West Midlands on 34%. The decline in the East Midlands has been limited to 5.1%.

Looking at the most recent five-year period, growth was achieved in the North West (2.9%) and the South West (0.9%). All other regions have seen some shrinkage, the largest being in the North East and the West Midlands, each of which have seen a 16%+ decline.

Table 41: English Shire Areas – Patronage by Region since 2005/06
Millions of Passenger Journeys, excluding PTE patronage

Year	North East	North West	Yorkshire & Humber	West Midlands	East Midlands	Eastern	South East	South West	Total
2005	77	129	43	98	200	167	283	170	1,168
2006	73	128	49	94	202	168	289	171	1,173
2007	73	134	49	90	202	183	310	188	1,229
2008	70	133	60	88	215	189	319	196	1,270
2009	69	134	62	87	216	194	328	205	1,294
2010	64	135	65	80	218	189	332	202	1,284
2011	64	136	66	74	215	190	338	206	1,290
2012	60	137	62	80	213	190	345	209	1,295
2013	51	138	51	79	207	186	345	202	1,259
2014	54	139	54	82	205	190	355	210	1,289
2015	52	140	48	78	203	187	355	216	1,278
2016	51	141	56	75	198	182	353	217	1,274
2017	47	142	70	70	195	188	356	220	1,290
2018	47	143	77	69	190	175	350	216	1,267
2019	44	144	75	65	190	177	349	218	1,261
% changes									
Since 2004/05	-43.3%	11.6%	74.8%	-34.0%	-5.1%	5.8%	23.1%	27.7%	7.9%
Last five years	-16.3%	2.9%	56.3%	-16.5%	-6.2%	-5.1%	-1.9%	0.9%	-1.3%
Last Year	-7.8%	0.7%	-2.4%	-5.5%	0.1%	1.4%	-0.5%	0.5%	-0.5%

Source: PTIS Analysis of DfT Annual Bus Statistics, Sheets BUS0108 and 0109.

The trends in each region can be seen in Figure 9-2, which shows an index of patronage (2004/05=100).

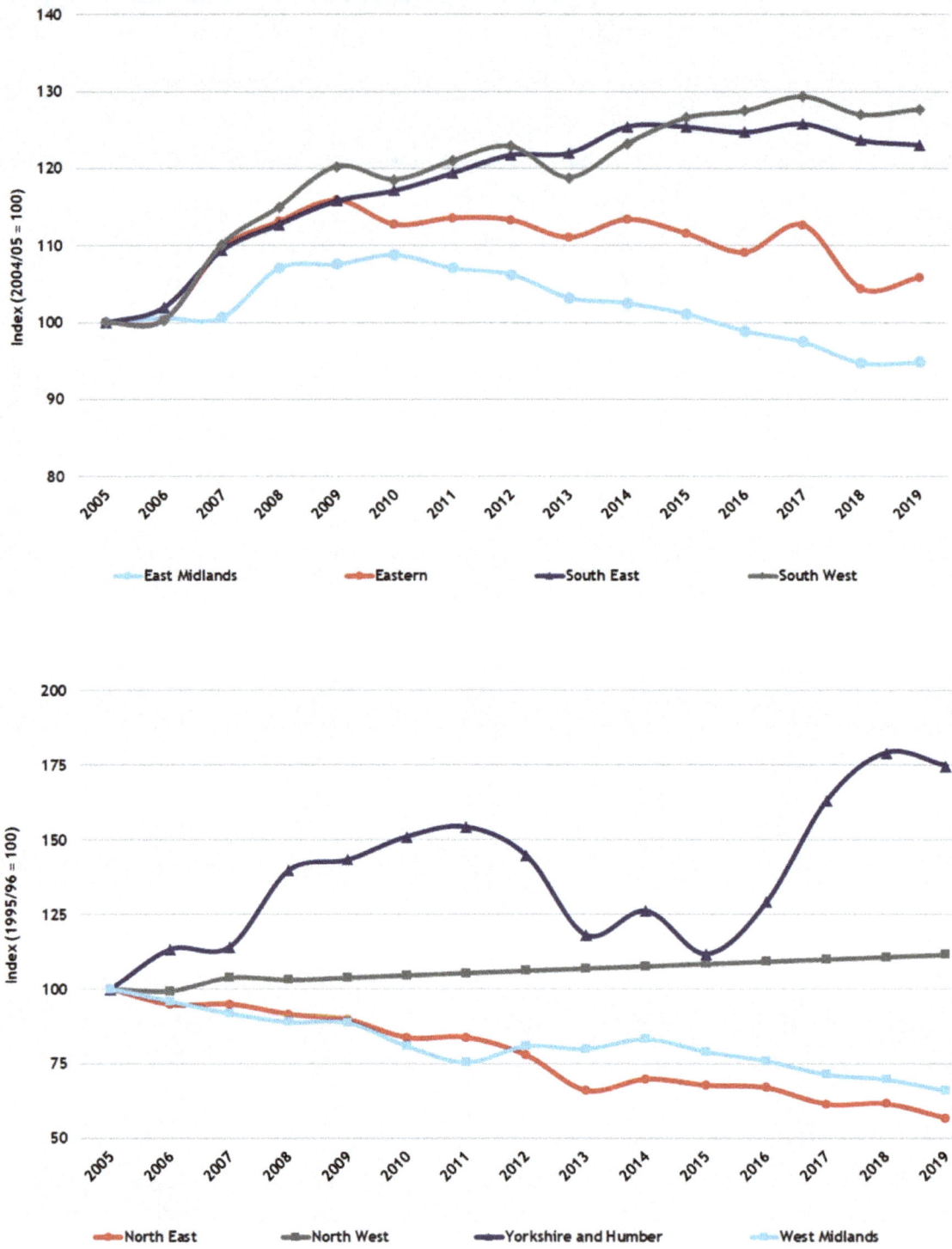

Figure 9-2: Bus Patronage Indices for Shire Areas, by region

9.4 Benchmarking Demand

As with the PTE areas, the benchmarking of demand can be achieved by considering the number of bus trips per head of the population.

The graph at Figure 9-3 shows the position in 2007/08, whilst Table 42 below provides a series of snapshots of the figures for the shire areas by region.

Figure 9-3: Bus Ridership per Head, Shire Areas by Region 2009 and 2019

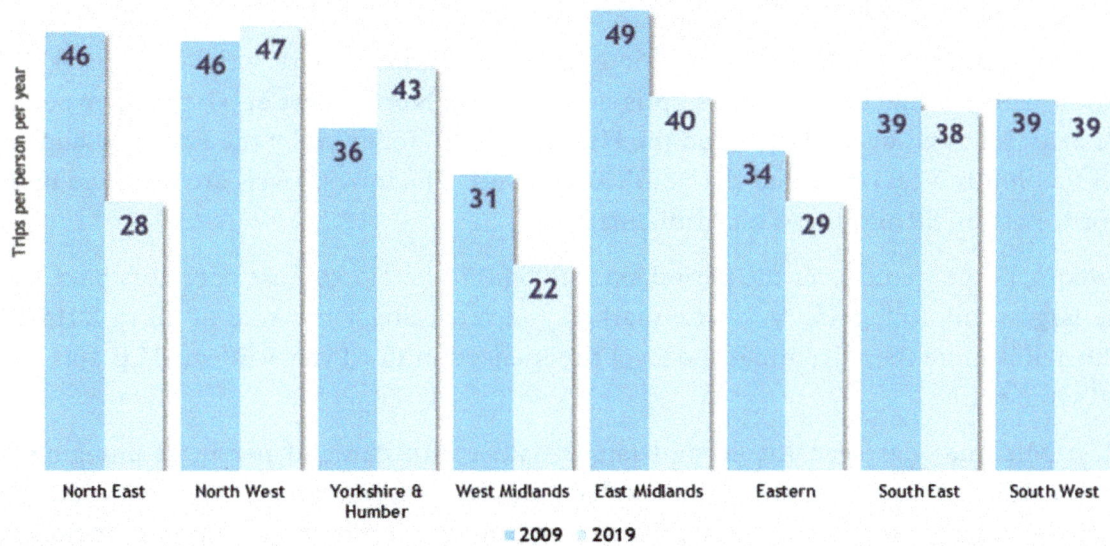

Table 42: Bus Ridership per Head of the Population, Shire Areas by region
Trips per person per year, excluding PTE Areas

Year	North East	North West	Yorkshire & Humber	West Midlands	East Midlands	Eastern	South East	South West	All Shire Areas
2005	52	44	26	36	47	30	35	34	38
2006	49	44	29	34	47	30	35	34	37
2007	50	45	29	33	46	33	38	37	38
2008	48	45	35	32	49	33	38	38	39
2009	46	46	36	31	49	34	39	39	40
2010	43	46	38	28	49	33	39	39	39
2011	43	46	39	27	48	33	40	39	39
2012	40	46	36	28	47	32	40	40	39
2013	34	46	30	27	45	31	40	38	37
2014	36	46	31	28	45	32	40	39	38
2015	35	47	28	27	44	31	40	40	37
2016	34	47	32	26	42	30	39	40	37
2017	31	47	40	24	41	31	39	40	37
2018	31	47	42	23	40	28	39	39	36
2019	28	47	43	22	40	29	38	39	36
% changes									
Since 2004/05	-45.7%	6.7%	64.4%	-39.1%	-15.5%	-6.3%	9.3%	14.9%	-6.5%
Last five	-17.6%	1.3%	53.9%	-18.7%	-9.4%	-7.9%	-4.6%	-2.3%	-4.0%
Last Year	-8.6%	0.3%	2.1%	-6.1%	-0.6%	0.9%	-1.0%	-0.2%	-0.8%

Source: PTIS Analysis of DfT Annual Bus Statistics

As can be seen, the average for the English shire areas is now 36 trips per person per year – less than half the average of 87 seen in the big urban areas covered by the PTEs, discussed in Chapter 8 above.

However, there are substantial variations between different parts of the country. The region with the highest bus ridership is currently the North West at 47 trips per person per year, followed by Yorkshire and the Humber on 43, followed by the East Midlands on 40, the South West on 39 and the South East on 40. The lowest levels are recorded in the North East on 28 and the West Midlands on 22.

Looking at the trends over the period since 2004/05, the North East operators have seen the largest fall, losing 45.7% of the market, and recording a trip rate of 28 in 2015/16. This fall is more than five times the level experienced in the Tyne & Wear PTE area over the same period.

West Midlands has seen the second largest fall – with demand per head plunging by 39.1%. It offers a particularly stark contrast with the West Midlands PTE area, which has some of the highest levels of per capita demand in our conurbations. Though the region is heavily oriented around the conurbation's major centres, there are nevertheless some substantial settlements in the shire areas, such as Warwick and Leamington, Telford, Stafford, Stoke and Worcester – which makes this statistic somewhat surprising.

The other areas to record a fall were the 15% fall in the East Midlands, still leaving the area with a trip rate of 40, and a 6.3% reduction in the East of England, taking it down to 29. The other shire areas – North West, Yorkshire and the Humber, Eastern, South East and the South West have all recorded increases in bus ridership since 2004/05 – the largest being almost 64% in Yorkshire.

In all cases, the impact of the introduction of free concessionary travel in 2006/07 can clearly be seen, pushing up overall ridership rates quite significantly, even if in some cases the full increase was not sustained into the following year. Since the onset of the recession, however, most areas have fallen back again.

9.5   Further Performance Differences

Even within regions, there are wide variations in performance between different authorities. Analysis of DfT statistics shows that patronage increases also occurred in 38 (43%) of the 88 local transport authorities (LTAs) in England during the year. Interestingly, 20 of those authorities can point to a record of growth in their areas across a five-year period, and 23 of them saw higher patronage in 2018/19 than in 2009/10. However, several – Bournemouth, Poole, Milton Keynes and North Somerset – reached a peak after 2010 but have fallen back from a peak since.

The LTAs within whose boundaries long term growth has been achieved are set out in Table 43 below. All of them are shire authorities and they are concentrated in the southern half of the country, being in the South East, South West and Eastern regions.

Table 43: LTAs with Bus Patronage Growth since 2009/10

Authority	Change Since 2009/10	Change in Last Five	Change
Bristol	52.2%	31.0%	11.4%
Central Bedfordshire	38.5%	16.9%	9.4%
Wokingham	38.2%	27.1%	18.1%
West Berkshire	37.7%	-0.1%	5.1%
Reading	36.1%	17.2%	4.1%
Bath & North East Somerset	29.9%	21.2%	23.8%
South Gloucestershire	28.2%	16.0%	-10.8%
Poole	26.5%	-7.4%	-8.0%
Thurrock	26.3%	3.6%	8.3%
North Somerset	26.0%	-3.0%	-27.3%
Brighton and Hove	22.2%	12.6%	1.7%
Oxfordshire	17.0%	-1.3%	3.1%
Luton	16.7%	32.9%	26.3%
Cornwall	15.6%	17.5%	12.8%
Southampton	10.3%	2.4%	-0.3%
West Sussex	8.7%	-2.3%	-1.8%
Torbay	7.3%	1.0%	3.1%
Bournemouth	5.4%	-11.0%	-7.2%
Milton Keynes	5.2%	-12.5%	-15.5%
Portsmouth	5.0%	3.4%	-5.3%
Nottingham	3.3%	0.7%	3.8%
Hampshire	3.1%	-3.0%	0.6%
Norfolk	3.1%	-1.4%	3.0%

Source: PTIS Analysis of DfT Annual Bus Statistics 2018/19, Sheet BUS 0109A

At the other end of the scale, there have been several LTA areas where the fall in demand is verging on the catastrophic. Thus, eight authorities show declines of more than ten per cent in 2018/19 alone. Over the five-year period, 15 authorities have recorded declines of more than 20 per cent - whilst the years since 2009/10 have seen 26 LTAs s with declines of more than 20 per cent. Of these, 13 are over 30% and two – Warrington and Stoke-on-Trent - show falls in excess of 40 per cent.

Table 44 contains a list of the shire authorities where a decline of more than 20 per cent has been seen since 2009/10. Most of the LTAs are in the midlands and the north of England, though some southern ones also appear in the list – the rural counties of Dorset and Somerset together with two unitary authorities in Bedford and Windsor & Maidenhead.

Table 44: LTAs with Bus Patronage Decline >20% since 2009/10

Authority	Change Since 2009/10	Change in Last Five	Change Last Year
Leicestershire	-20.0%	-5.2%	4.8%
Dorset	-20.5%	-17.6%	-1.1%
Lincolnshire	-21.1%	-19.4%	-4.5%
Cumbria	-21.6%	-13.0%	-0.5%
Leicester	-21.8%	-0.7%	-0.5%
East Riding of Yorkshire	-22.1%	-21.3%	-11.8%
Bedford	-24.9%	-5.9%	-6.2%
Shropshire	-25.6%	-8.6%	4.2%
Northumberland	-26.0%	-14.0%	0.6%
Staffordshire	-26.8%	-23.0%	-2.8%
Middlesbrough	-27.3%	-11.5%	1.2%
North Yorkshire	-28.2%	-23.5%	0.4%
North Lincolnshire	-28.9%	-26.9%	-12.4%
Blackburn with Darwen	-30.0%	-15.3%	-7.2%
Darlington	-30.4%	-11.4%	-4.3%
Blackpool	-30.7%	-19.9%	-5.0%
Lancashire	-31.7%	-16.4%	0.6%
Cheshire East	-32.4%	-28.3%	-7.6%
Worcestershire	-34.3%	-24.7%	-2.4%
Herefordshire	-34.7%	-27.0%	-3.0%
Somerset	-35.2%	-28.9%	-22.0%
Telford & Wrekin	-35.6%	-23.8%	-4.6%
Redcar & Cleveland	-36.1%	-12.0%	-4.6%
Windsor & Maidenhead	-38.2%	-24.8%	-6.5%
Stoke-on-Trent	-40.6%	-22.8%	-10.9%
Warrington	-48.5%	-29.7%	-5.9%

Source: PTIS Analysis of DfT Annual Bus Statistics 2018/19, Sheet BUS 0109A

9.6　Demographic & Economic Data

9.6.1　Population

Over the period since the 1981 census, the population of the English Shire areas has risen by around 20%. Since 2004/05, the growth has been 9.9%. The total stood at 35.0 million people in 2018, according to the ONS Mid-Year Population Estimates.

Increases have been seen in all regions over that period, with the South East and the East of England each seeing the largest growth at 11.9%, closely followed by the East Midlands on 11.6% and the South West on 10.5%. The South East is the largest region by some distance, with a population of 9.1 million in 2018.

The East of England (or Eastern) region is the next largest on 6.2 million, closely followed by the South West on 6.0 million.

At the other end of the scale, the slowest growth has been seen in the North East on 3.8%.

Figure 9-4: Population Change, Shire Areas of English Regions, 2005-2018

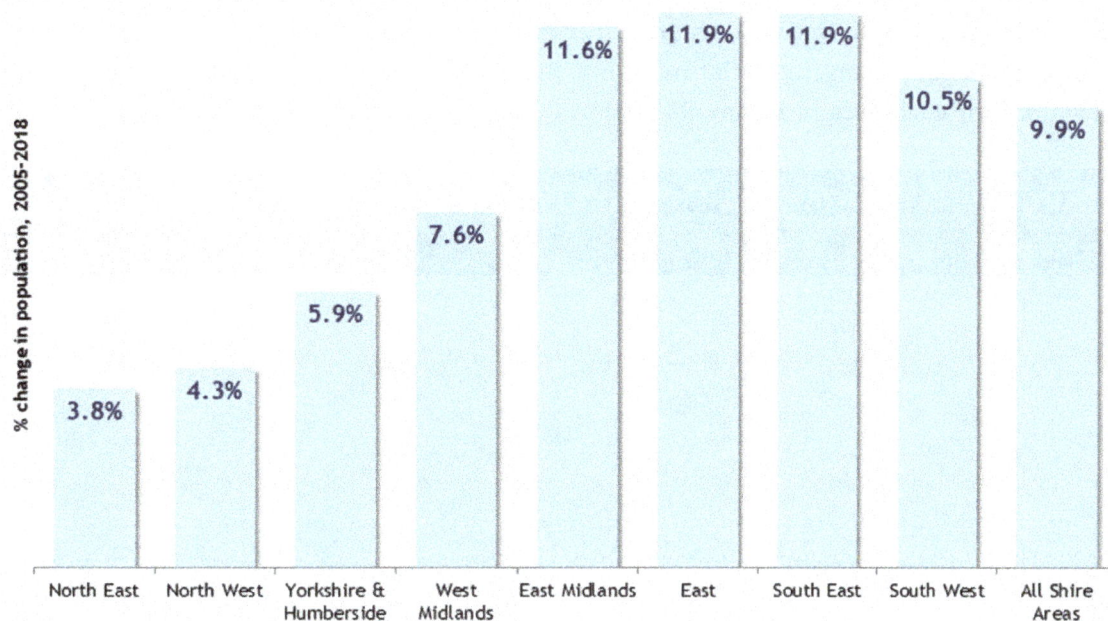

9.6.2   Employment

The ONS Labour Force Survey enables us to track the change in employment across the country, and the picture since 2005 has been of overall growth in all areas, albeit with some falls during the immediate aftermath of the financial crisis in 2008 and again in 2012.

The changes over the period are illustrated in the chart at

Figure 9-5 below. Across all shire areas since 2005, around 1½ million extra jobs have been created – growth of 8.1%. However, there is considerable variation between regions, ranging from growth of over 13% in the three south regions, and just 2.7% in Yorkshire outside the conurbations.

Figure 9-5: Employment Growth in the English Shires, 2005-2019

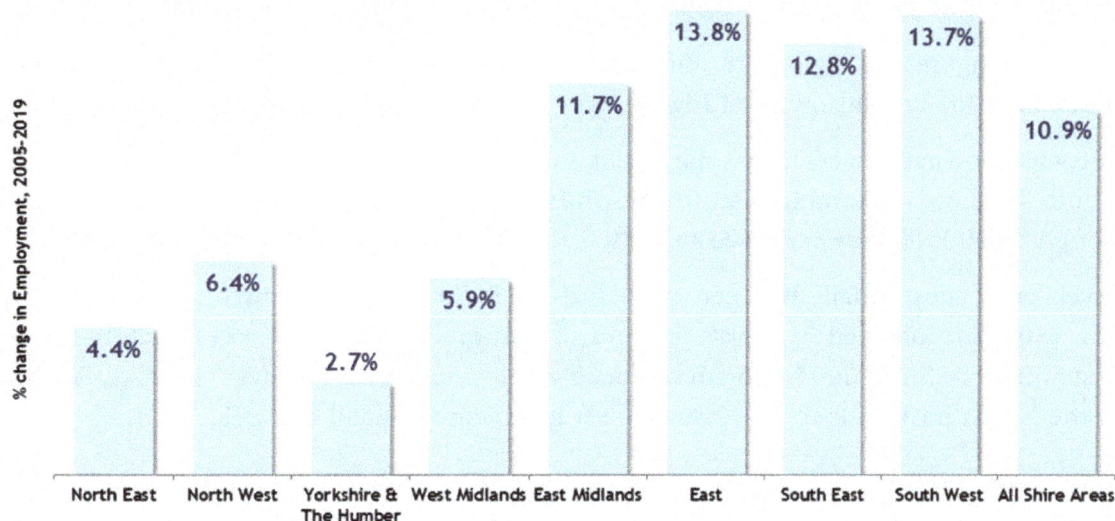

Source: NOMIS Database (ONS/University of Durham).

9.6.3   Economic Growth

The ONS figures on Gross Value Added (GVA) for each area shows a picture of the economic growth achieved in each area over the period between 2005 and 2018. The real-term change (adjusting for inflation using the GDP Deflator) over the whole period is measured for each area in Figure 9-6 below.

Figure 9-6: GVA Growth at Constant Prices, English Shires, 2005-2019
% change by region

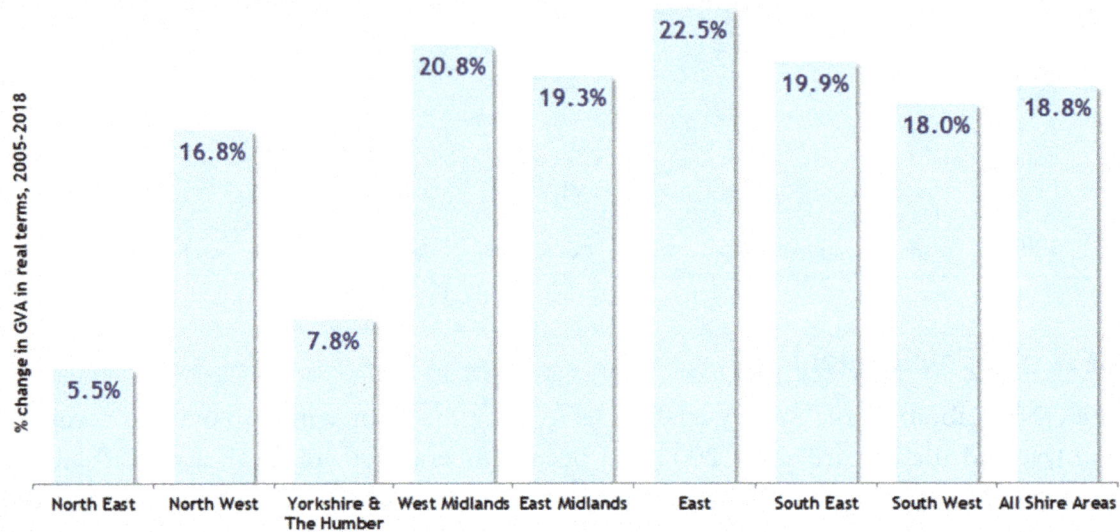

Source: ONS

As can be seen, the shire areas have seen real growth of 18.8% during the period, with results varying across the regions. The shire areas of the North East have seen the lowest growth, at 5.5%, whilst the non-urban parts of Yorkshire and the Humber saw growth limited 7.8%. All other regions grew by more than 15% with the East leading the way on 22.5%.

9.6.4   Car Ownership

The 2018 figures by region are illustrated in Figure 9-7, whilst the changes since both 1986 and 2005 are illustrated in Figure 9-8 . The details are contained in Table 45.

The gap in ownership levels has narrowed – in 2018, the gap was 19.7% between highest (South West on 574) and lowest (the North East, 461). In 1986 that gap was almost 43%, though by 2005, it had narrowed to 22%.

Levels grew substantially between 1986 and 2018. The highest growth was in the North East (80.6%), followed by Yorkshire and the Humber (69.2%). Lowest was the South East at 27.9%. Since 2005, growth has been greatest in the South West (11.4%), followed by the North East (8.4%). The North West has recorded a fall of 8.2%.

Figure 9-7: Car Ownership per 1,000 population, Shire Areas, 2018

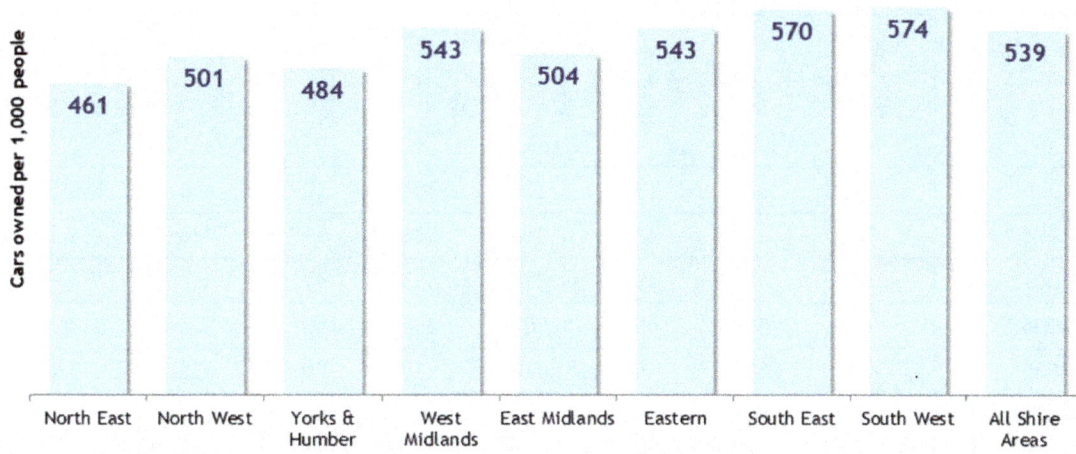

Cars owned per 1,000 people

North East	461
North West	501
Yorks & Humber	484
West Midlands	543
East Midlands	504
Eastern	543
South East	570
South West	574
All Shire Areas	539

Figure 9-8: Change in Car Ownership since 1986 and 2005, Shire Areas

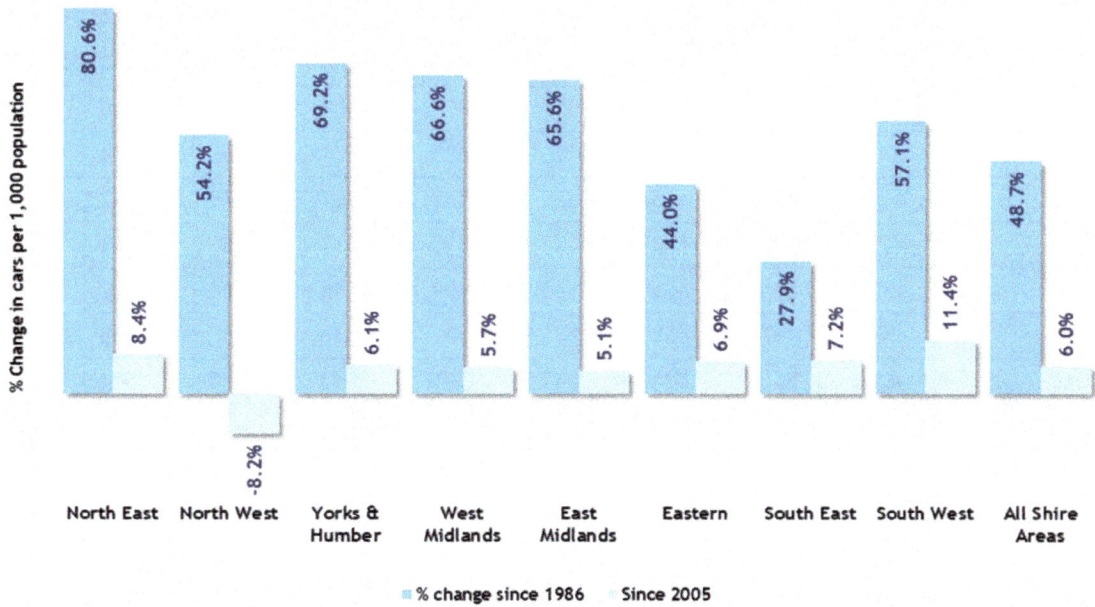

% Change in cars per 1,000 population

Region	% change since 1986	Since 2005
North East	80.6%	8.4%
North West	54.2%	-8.2%
Yorks & Humber	69.2%	6.1%
West Midlands	66.6%	5.7%
East Midlands	65.6%	5.1%
Eastern	44.0%	6.9%
South East	27.9%	7.2%
South West	57.1%	11.4%
All Shire Areas	48.7%	6.0%

■ % change since 1986   Since 2005

95

Table 45: Car Ownership per 1,000 people, Shire Areas

Region	North East	North West	Yorks & Humber	West Midlands	East Midlands	Eastern	South East	South West	All Shire Areas
1986	255	325	286	326	304	377	446	365	363
1994	302	372	322	418	376	398	426	414	395
2001	348	413	399	453	423	471	484	473	452
2004	415	505	451	507	468	502	524	510	498
2005	425	546	456	514	479	507	532	515	509
2006	426	556	448	517	474	503	531	511	507
2007	430	552	452	526	480	504	534	513	510
2008	435	553	454	528	484	503	536	516	512
2009	435	515	453	528	482	502	540	520	510
2010	432	492	452	524	481	501	540	521	507
2011	434	475	456	515	478	504	533	527	504
2012	435	475	456	517	478	504	537	529	506
2013	438	478	460	520	483	521	543	534	513
2014	444	483	466	525	489	527	549	542	519
2015	445	488	473	534	495	537	556	553	527
2016	455	494	479	540	499	545	566	562	535
2017	458	497	458	541	501	546	567	568	536
2018	461	501	484	543	504	543	570	574	539
% changes									
Since 1986	80.6%	54.2%	69.2%	66.6%	65.6%	44.0%	27.9%	57.1%	48.7%
Since 1994	52.7%	34.9%	50.4%	29.8%	34.2%	36.4%	33.8%	38.6%	36.5%
Since 2001	32.4%	21.5%	21.4%	19.9%	19.0%	15.1%	17.8%	21.4%	19.4%
Since 2005	5.9%	-9.3%	6.6%	2.7%	4.2%	7.9%	6.3%	11.3%	5.4%

Source: PTIS analysis of Mid-Year Population Estimates (Office for National Statistics) and Vehicle Licensing Statistics (Department for Transport).

9.7 Conclusions

As can be seen from the data in this chapter, the conditions for growth in bus patronage – in terms of growth in population, employment and the wider economy – have existed to some extent in these areas throughout the last fifteen years. Though car ownership rates have continued to grow, the percentage changes have been moderate, and the growth was interrupted by the hiatus following the 2008 financial crisis – not yet recovering in one region.

Data from the National Travel Survey suggests that the number of households not owning a car has not fallen significantly over the period since 2005, moving from 23% to 21% in urban cities and towns. Rural areas have seen the proportion of no-car households falling from 16% in 2005/06 down to 14% in 2017/18. However, this fall took place early in the period, and the 14% figure has remained remarkably constant ever since.

Thus, as with the PTE areas discussed earlier, we must look elsewhere for the reasons for the significant decline in patronage that has taken place across these areas. This is discussed further in Chapter 13 below.

Chapter 10: Scotland

10.1   Overall Trends

It was in 1975 that the publication of separate bus patronage figures for Scotland began. In that year, Scotland's buses carried 891 million passengers, just under 12% of the GB total of 7,524 million.

From that figure of 891 million, the decline carried on, largely unchecked, reaching a low point in the year before devolution, 1998/99, with just 424 million passengers. It then began a long period of recovery and indeed almost reached the 500 million mark in 2007/08 immediately before the full effects of the financial crisis began to be felt.

The figures since 1975 are illustrated graphically at Figure 10-1 below.

Figure 10-1: Historic Trends in Bus Patronage, Scotland since 1975

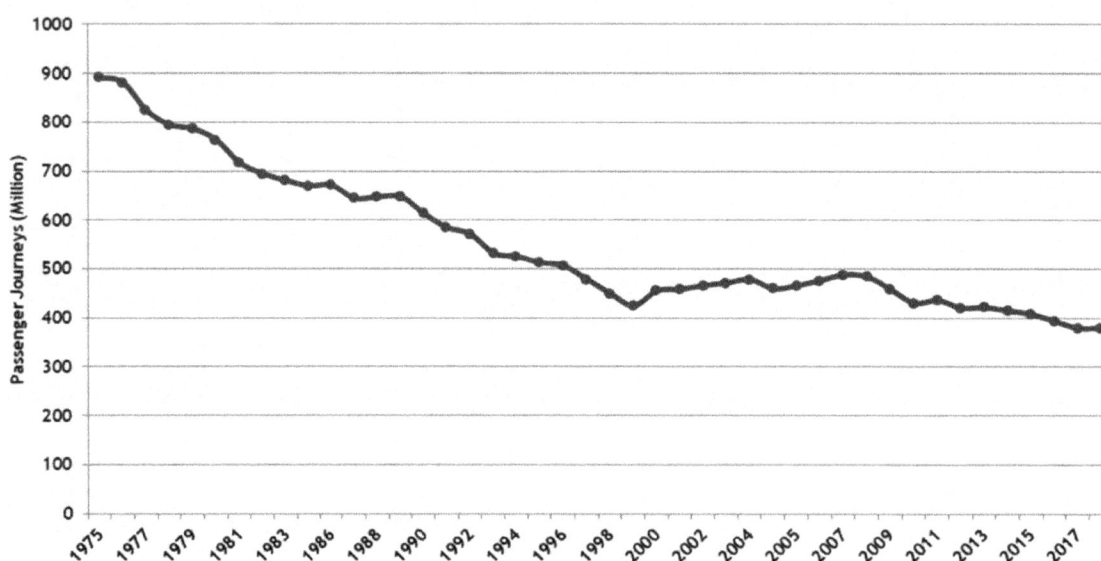

Source: Annual Bus Statistics, DfT

The key market figures for each year since 2004/05 are set out in the table at Table 46 below. Passenger journey numbers hit a pre-recession peak of 484 million journeys in 2008/09, but this was followed by the sharp falls during and after the recession, so that the 2018/19 figure was more than 100 million lower, at 380 million. During the decade after the crisis, there were only two years of growth, in 2011/12 and 2018/19.

Passenger kilometres travelled have fallen in line with the number of journeys. Revenue has increased in real terms over the period since 2004/05 but reached a peak in 2016/17 and has fallen back since. Meanwhile, service supply has also fallen as tendered service budgets and commercial networks have both come under pressure.

Indices of these measures, based on 2004/05=100, are shown in Figure 10-2 below.

Table 46: Key Traffic Statistics for Bus Services in Scotland

Year to 31 March	Passenger Journeys (million)	Passenger Kilometres (Millions)	Passenger Revenue £m (2015/16 Prices)	Kilometres Run
2005	459	3,221	478	359
2006	465	3,451	502	374
2007	476	3,530	559	385
2008	487	3,596	586	397
2009	484	3,763	611	386
2010	458	3,753	602	377
2011	430	3,477	583	346
2012	436	3,200	594	338
2013	420	2,938	581	327
2014	421	2,943	590	332
2015	414	2,895	595	328
2016	407	2,844	595	331
2017	393	2,747	607	327
2018	379	2,714	585	333
2019	380	2,654	574	334
% changes				
Since 2005	-17.3%	-17.6%	20.1%	-7.2%
Last Five Years	-8.3%	-8.3%	-3.5%	1.6%
Last Year	0.1%	-2.2%	-1.9%	0.2%

Figure 10-2: Key Traffic Indices for Bus Services in Scotland

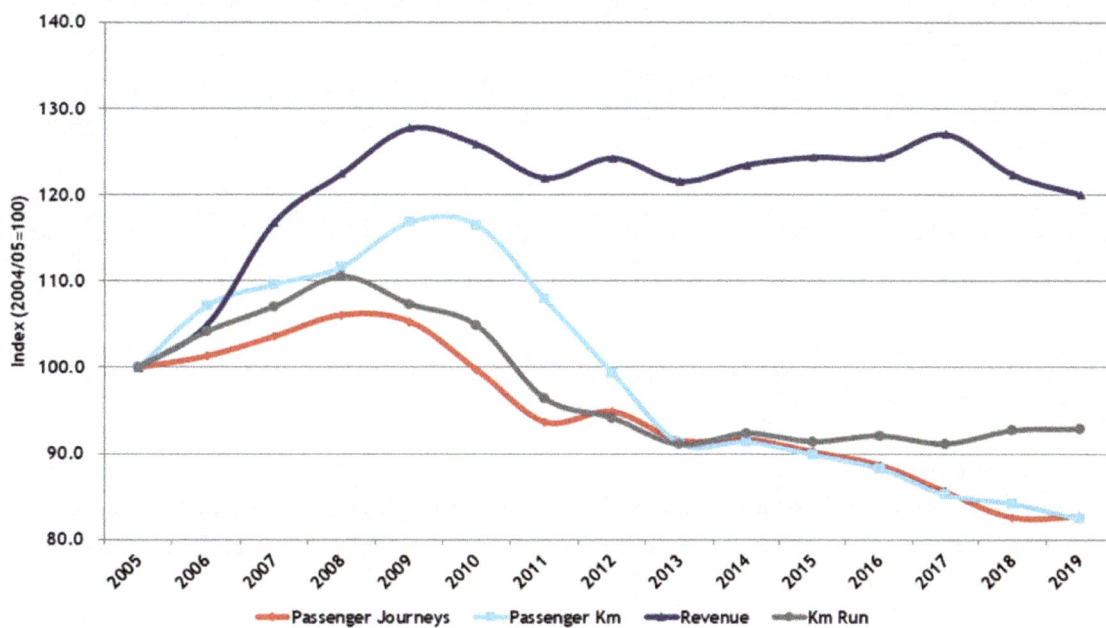

10.2   Commercial Performance

Table 47 below shows the analysis of market trends derived from the traffic statistics. Over the period since 2004/05, average journey length initially grew, reaching over 8 kilometres when patronage was at its peak just before the recession. It then fell back again to around 7 and has remained pretty constant ever since. Revenue per passenger and per passenger kilometre show significant real-term gains of around 45%. However, average loads have fallen by around 11%.

Over the most recent five-year period, average journey length has been unchanged, but there have once again been increases in real-term yields of around 5%. Average bus load has fallen by 2.4% to a figure of 7.96 – falling below eight for the first time ever.

Table 47: Scottish Bus Statistics - Market Analysis

Year to 31 March	Average Journey (kilometres)	Average Fare (£, 2018/19 prices)	Yield (£, 2018/19 Prices	Average Load (Passenger Kms per Kilometre run)
2005	7.01	1.041	0.148	8.96
2006	7.42	1.078	0.145	9.22
2007	7.42	1.174	0.158	9.18
2008	7.38	1.202	0.163	9.06
2009	7.78	1.263	0.162	9.76
2010	8.19	1.314	0.160	9.96
2011	8.08	1.355	0.168	10.04
2012	7.35	1.364	0.186	9.47
2013	6.99	1.383	0.198	8.98
2014	6.99	1.402	0.201	8.87
2015	6.99	1.435	0.205	8.82
2016	6.98	1.460	0.209	8.60
2017	6.98	1.545	0.221	8.39
2018	7.15	1.542	0.216	8.15
2019	6.99	1.511	0.216	7.96
% changes				
Since 2005	-0.4%	45.2%	45.7%	-11.3%
Last Five Years	0.0%	5.3%	5.3%	-9.8%
Last Year	-2.3%	-2.0%	0.3%	-2.4%

10.3   Regional Trends

Data on the performance of bus and coach services in Scotland are available in the Scottish Transport Statistics, published by the Transport Scotland. They publish disaggregated bus patronage statistics based on four areas. The available data series commences in 2004/05 and is shown in Table 48 below.

The areas are:

- **North East, Tayside and Central**, including Perth and Kinross, Stirling, Aberdeen City, Aberdeenshire, Angus and Dundee City

- **Highlands, Islands and Shetland**, covering Na h-Eileanan Siar [Western Isles], Highland, Moray, Orkney Islands, Shetland Islands and Argyll & Bute

- **South East** including Clackmannanshire, East Lothian, Falkirk, Fife, Midlothian, Scottish Borders, Edinburgh City and West Lothian

- **South West and Strathclyde**, incorporating Dumfries & Galloway, East Ayrshire, East Dunbartonshire, East Renfrewshire, Inverclyde, North Ayrshire, South Ayrshire, South Lanarkshire, Renfrewshire, West Dunbartonshire, Glasgow City and North Lanarkshire

Table 48: Bus Patronage (millions) in Scotland since 2005, by Region

Year to 31 March	North East, Tayside and Central	Highlands, Islands and Shetland	South East	South West and Strathclyde	Scotland
2005	68.0	11.0	161.0	220.0	461.0
2006	67.0	11.0	164.0	224.0	468.0
2007	65.0	15.0	174.0	223.0	476.0
2008	68.0	14.0	174.0	232.0	487.0
2009	66.0	14.0	170.0	234.0	484.0
2010	61.0	14.0	162.0	219.0	458.0
2011	61.3	14.4	161.6	193.0	430.2
2012	62.6	13.0	165.9	194.1	435.7
2013	61.0	13.6	162.2	183.6	420.3
2014	61.8	12.6	164.3	182.3	421.0
2015	62.9	12.4	163.7	175.2	414.3
2016	59.5	13.4	161.8	172.3	406.9
2017	58.3	11.7	158.9	165.5	394.3
2018	55.3	10.9	157.2	164.9	388.3
2019	53.1	10.4	157.0	159.2	379.8
Since 2005	-21.9%	-5.5%	-2.5%	-27.6%	-17.6%
Last Five	-15.6%	-16.3%	-4.1%	-9.1%	-8.3%
Last year	-3.8%	-4.6%	-0.1%	-3.5%	-2.2%

Source: Scottish Transport Statistics

Two of the regions have seen significant reductions in demand over the period since 2004/05, with the largest occurring in the South West and Strathclyde (27.6%). The North East, Tayside and Central region has also seen some loss, but at a lower percentage of 21.9%. In contrast, the South East has seen shrinkage of 2.5% over the period, and at least some of that may be attributable to the opening of Edinburgh tram in 2014/15.

Services in the two southernmost regions account for the lion's share of Scottish demand. The 316 million journeys made in 2018/19 represented 83.3% of demand for the whole country, even though the two regions only account for 70.5% of the population.

Both areas have experienced considerable variation since 2004/05. For example, in Strathclyde and the South West, patronage grew in the middle part of the last decade, reaching a peak of 234 million trips in 2007/08. Since then there has been a marked decline, with the most recent year available showing just 159 million trips.

Across to the east, there was also substantial growth, peaking in 2008 and 2009 at 174 million trips. There has been a fall since, but by no means as great as in the South West.

The graph at Figure 10-3 below illustrates the trends.

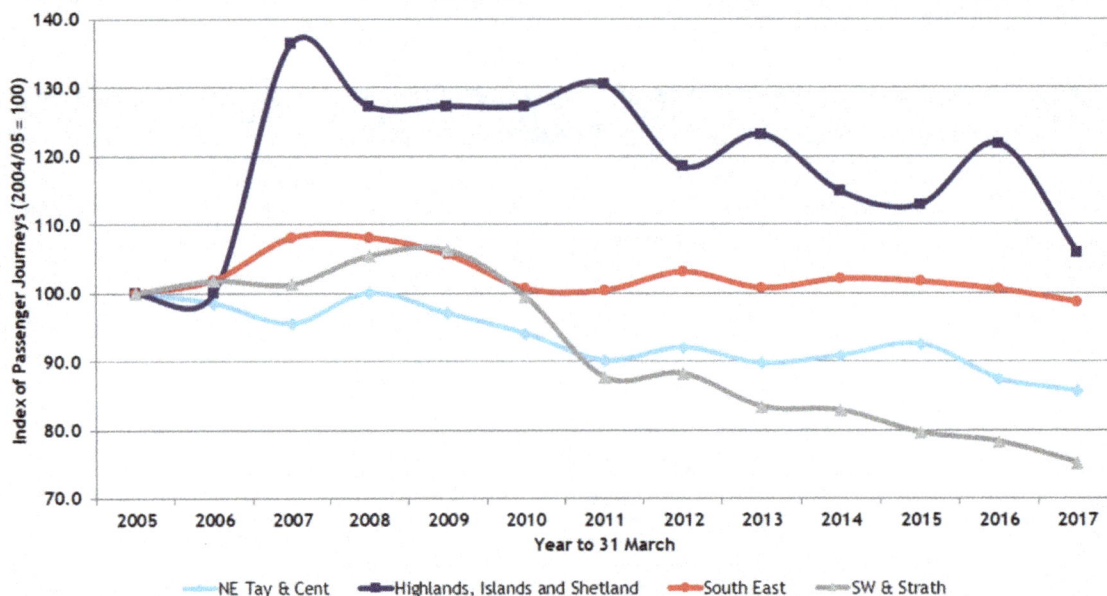

Figure 10-3: Bus Patronage Indices for Scotland, by Region

10.4   Demand Benchmarking in Scotland

The *per capita* analysis for the Scottish regions is shown in Table 49 below and illustrated graphically in Figure 10-4, which gives a snapshot in 2008/09 and 2018/19.

It is notable that ridership levels for the two regions covering Scotland's two biggest cities are very different. The south eastern area including Edinburgh has a much higher ridership level at 113 trips per person per year – amongst the highest in the UK, in fact. Meanwhile, the South Western area reached a peak of 97.8 trips per person per year in 2008/09 but has since fallen back sharply, standing at 64.6 in 2018/19.

The original disparity between the two may be associated with the much more intensive urban rail network operated in the Strathclyde area. However, the fall in ridership levels in south west Scotland is very large and merits further investigation. Rising congestion levels and falling bus speeds have reportedly affected the city of Glasgow very badly, and this is held to be at least part of the explanation.

Table 49: Bus Ridership per Head of Population, Scotland by region

Year to 31 March	North East, Tayside and Central	Highlands, Islands and Shetland	South East	South West and Strathclyde	Scotland
2005	56.6	43.1	127.9	92.4	90.5
2006	55.5	42.8	129.1	94.1	91.5
2007	53.3	57.8	135.8	93.5	92.5
2008	55.4	53.6	134.7	97.2	94.2
2009	53.4	53.3	130.5	97.8	93.2
2010	51.3	53.0	123.2	91.4	87.7
2011	48.0	51.7	122.8	79.5	81.2
2012	48.7	46.9	125.4	80.0	82.0
2013	47.3	48.8	121.9	75.6	78.9
2014	47.6	45.4	122.7	75.0	78.7
2015	48.2	44.5	121.2	71.9	77.1
2016	48.1	51.0	124.2	72.0	78.3
2017	46.7	44.1	120.8	69.1	75.5
2018	42.3	38.8	113.5	66.9	71.4
2019	40.7	37.0	113.4	64.6	69.8
% changes					
Since 2005	-13.5%	-0.7%	-5.0%	-22.8%	-14.8%
Last Five	1.0%	-20.5%	-3.7%	-11.6%	-6.8%
Last year	1.0%	-8.0%	-1.4%	-5.2%	-3.0%

Figure 10-4: Bus Ridership per Head, Scotland by region 2009 & 2019

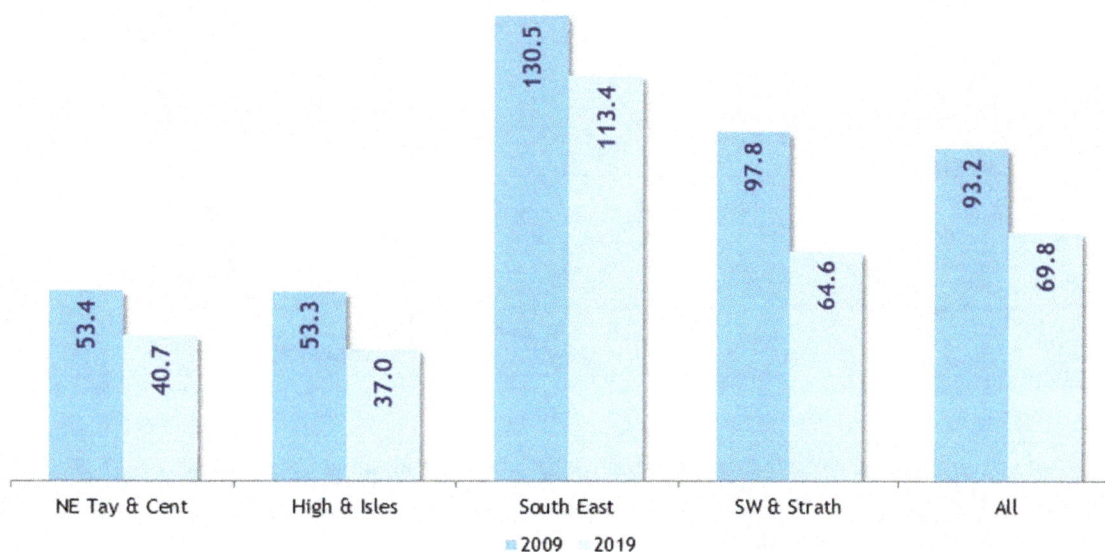

103

10.5 Demographic & Economic Data

10.5.1 Population

Although Scotland's population edged towards and just beyond the five million mark just before the Second World War, it was not until 1947 that this significant figure was consistently exceeded. It has stayed above that line ever since, reaching a peak in 1974 at 5.24 million. Thereafter there was a gentle decline in most years until the low point was reached in 2001, when the estimate stood at 5.064 million. It has since staged a recovery and passed the previous record high in 2010. The 2015 estimate was 5.373 million.

The population of the country is very much concentrated to the central belt linking Glasgow and Edinburgh, with the two transport partnerships covering these areas accounting for over 70% of the population. Population densities[6] vary from amongst the highest in the UK outside London in Glasgow (3,579 people per square kilometre) to the lowest (Eilean Siar on 8.8 and Highlands on 9.18).

The long-term changes are illustrated in the graph at Figure 10-5 below. The population totals for each of the regions discussed earlier in the chapter are shown in Table 50. The densities of the regions are illustrated in

Figure 10-6.

Figure 10-5: Population Trends in Scotland since 1961

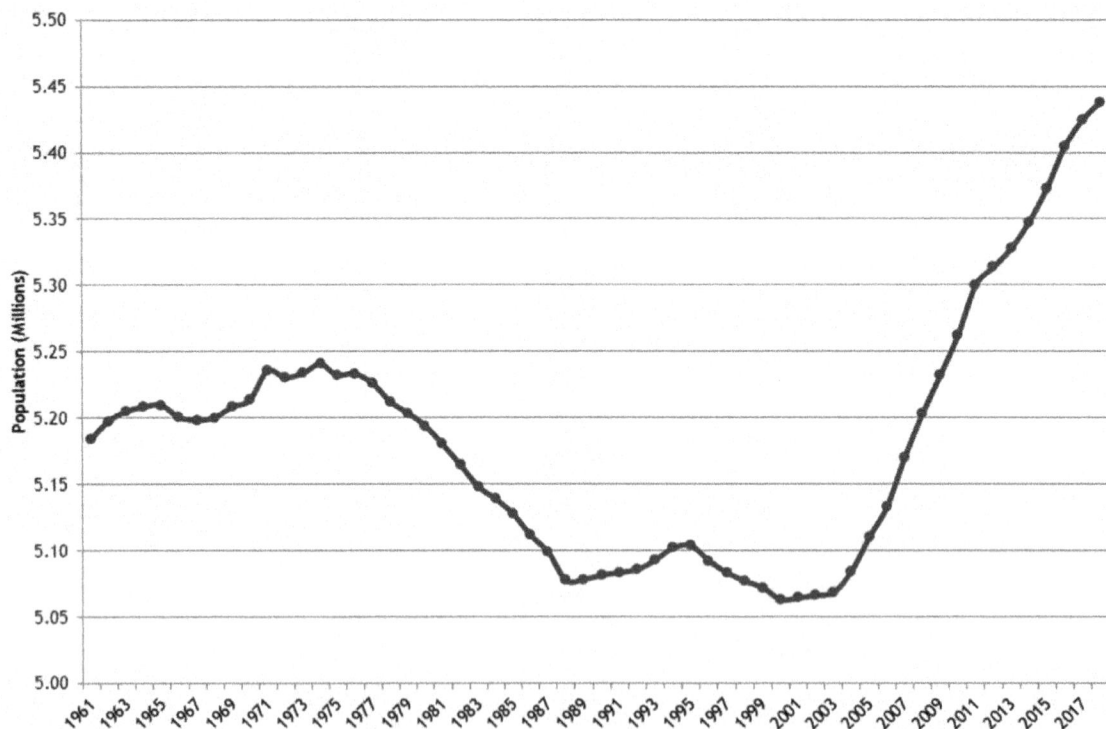

Source: Population Estimates for the UK and its constituent countries, 1838-2015 and Mid-Year Population Estimates. Office for National Statistics

Over the period since 1986, the Highlands and Islands have seen growth of 14.2%, whilst the south east areas around and to the south of Edinburgh have seen growth of 15.7%. Double digit growth of 11% has also been seen in the North East, Tayside and Central

---

[6] Office of National Statistics Mid-Year Population Estimates 2018

region, though the major downturn in the oil industry in the north east has seen some recent falls. The South West, covering Glasgow and Strathclyde, has been broadly unchanged, with falls in 2004 and 2005 counteracted in the years since the recession.

Table 50: Population Trends in Scotland since 1986, by region

Year	North East, Tayside and Central	Highlands, Islands and Shetland	South East	South West and Strathclyde	Total for Scotland
1986	1,166	240	1,187	2,518	5,112
1991	1,177	246	1,197	2,464	5,083
1995	1,205	251	1,214	2,434	5,104
1999	1,200	251	1,225	2,397	5,072
2001	1,194	250	1,236	2,384	5,064
2004	1,194	253	1,252	2,380	5,078
2005	1,200	255	1,259	2,380	5,095
2006	1,208	257	1,270	2,382	5,117
2007	1,218	259	1,282	2,385	5,145
2008	1,227	261	1,292	2,388	5,169
2009	1,236	263	1,303	2,392	5,194
2010	1,247	264	1,315	2,396	5,222
2011	1,278	277	1,316	2,429	5,300
2012	1,284	278	1,324	2,428	5,314
2013	1,291	278	1,330	2,428	5,328
2014	1,298	278	1,339	2,432	5,348
2015	1,306	279	1,350	2,439	5,373
% changes					
Since 1986	10.9%	13.5%	12.8%	-1.0%	5.7%
Since 2005	8.8%	9.3%	7.2%	2.5%	5.5%
Last Five	2.2%	0.6%	2.6%	0.4%	1.4%
Last year	0.5%	0.4%	0.8%	0.3%	0.5%

Source: PTIS analysis of Mid-Year Population Estimates, Office for National Statistics.

Figure 10-6: Population Densities by Scottish Region, 2018

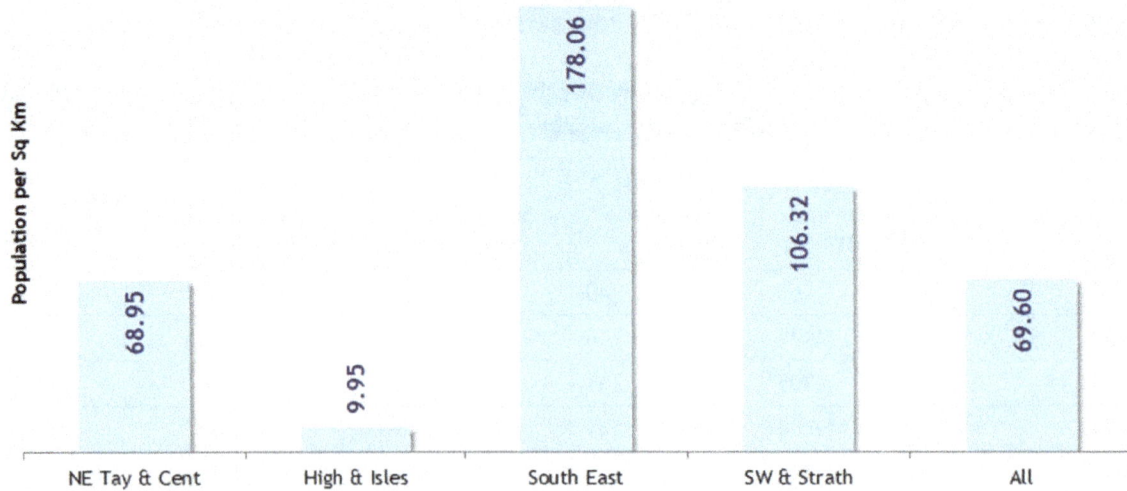

Population per Sq Km

Region	Population per Sq Km
NE Tay & Cent	68.95
High & Isles	9.95
South East	178.06
SW & Strath	106.32
All	69.60

Source: PTIS Analysis of Mid-Year Population Estimates, ONS

10.5.2   Employment

The data for employment in Scotland over the last decade shows growth of just 1.1% overall. The ONS NOMIS database shows that, during the run-up to the recession, the number of people economically active and in employment peaked at 2.78 million in March 2008.

Within two years some 180,000 jobs had disappeared. Since 2010, however, there has been a gradual recovery. The 2008 peak figure was eventually exceeded again in 2014, since when numbers have kept on growing, though there was a setback in 2016 – at least in part because of the effect of the sharp falls in the oil price just as other sectors were recovering. In March 2019, the total stood at 2.70 million.

Figure 10-7: Economically Active and in Employment, Scotland since 2004

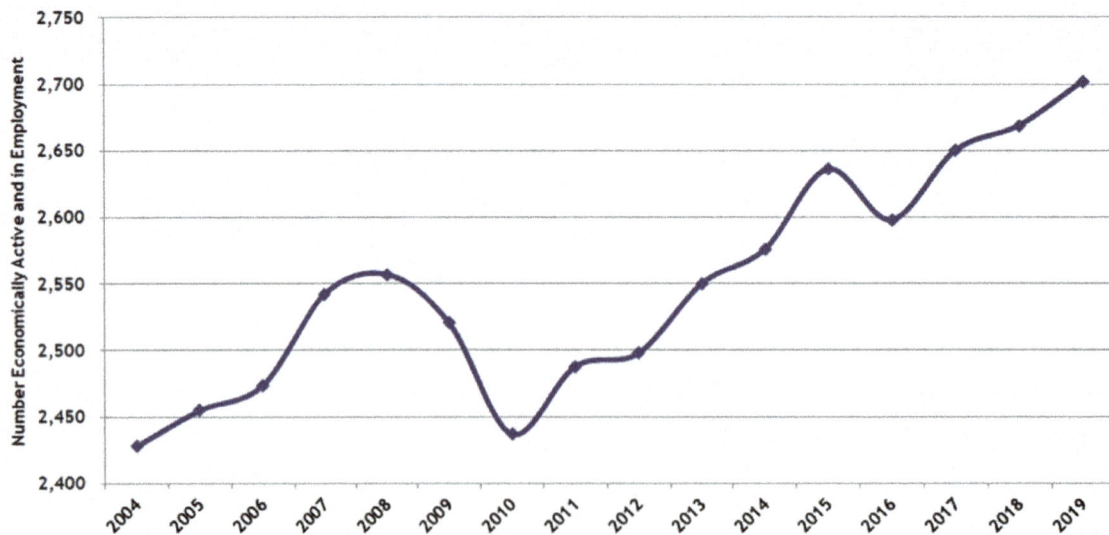

Source: Office for National Statistics, NOMIS database

10.5.3   Economic Growth

In contrast to its record on employment, Scotland's economic growth as measured by Gross Value Added at constant (2018) prices shows a relatively strong performance since 2005, with growth of 18% over the period. Again, there were setbacks during the recession, but these tended to be much lower than in many parts of England, with the result that output has recovered strongly. The effect of the collapse in the oil price in 2015 and 2016 on the Scottish economy can clearly be seen.

Figure 10-8: Movements in GVA at constant (2018) prices, Scotland

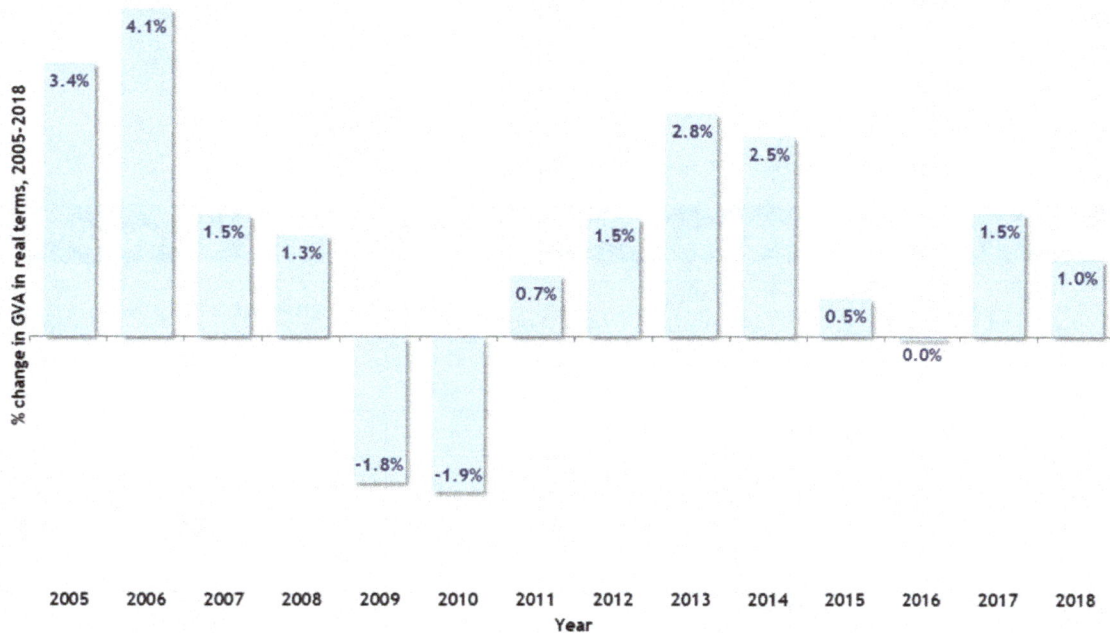

Source: Office for National Statistics. Adjusted for inflation using 2018 GDP Deflator

10.5.4   Car Ownership

Over the period since 1986, car ownership has grown more quickly in Scotland than in any other part of the UK, rising by 87.8%. In 1986, the national figure was 243 cars per 1,000 people. By 1994, this had grown to 311, reaching 360 by 2001. In 2004, the level reached 409, and it peaked just prior to the recession at 433 in 2008/09. It fell back in 2010/11 but started to grow again in 2011/12 and has continued to do so, reaching 457 in 2018.

As with England, there are variations between different parts of the country. More northerly and less sparsely populated areas have higher car ownership levels, whilst the more urban parts have lower levels. Disaggregated data from 2009 onwards has been analysed into the four broad regions used for bus patronage data. The figures for 2018 are illustrated in Figure 10-9, whilst the figures for each year are shown in Table 51 below.

Figure 10-9: Car Ownership per 1,000 people, Scottish Regions, 2018

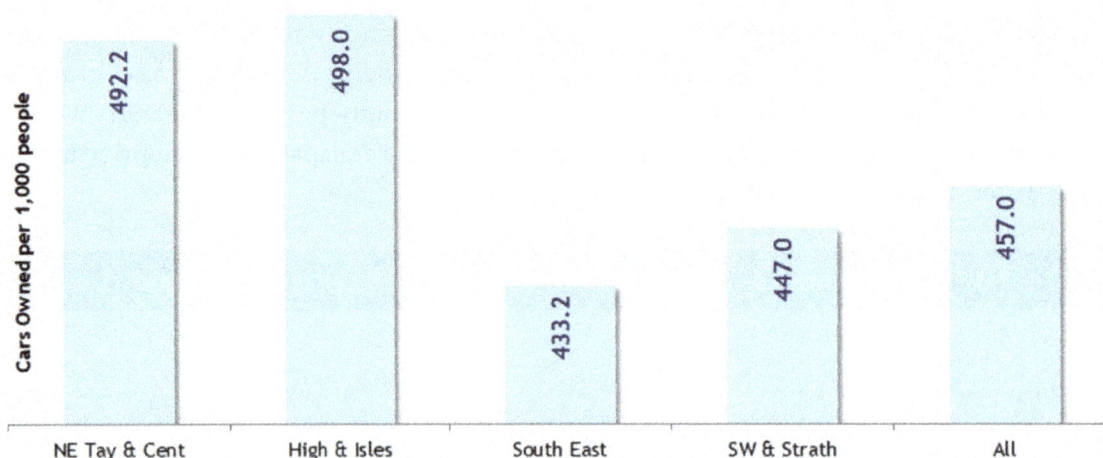

Source: *Vehicle Registration Statistics, DfT and Mid-Year Population Estimates, ONS*

Table 51: Car Ownership per 1,000 people, Scottish Regions since 2009

Year	North East, Tayside & Central	Highlands, Islands & Shetland	South East	South West & Strathclyde	All Scotland
2009	474.3	485.0	414.2	416.0	432.9
2010	471.9	484.2	412.0	415.5	431.5
2011	464.2	462.8	414.1	410.6	427.1
2012	467.2	466.5	414.7	414.3	429.9
2013	473.3	473.7	417.9	420.0	435.2
2014	479.6	480.8	423.2	429.9	442.9
2015	479.7	484.7	426.0	433.5	445.5
2016	483.9	490.8	429.6	438.8	450.1
2017	488.4	495.3	432.6	442.5	453.8
2018	492.2	498.0	433.2	447.0	457.0

Source: *Vehicle Registration Statistics, DfT and Mid-Year Population Estimates, ONS*

Data from the Scottish Household Survey shows that the proportion of households without access to a car across the country in 2018 was 28.6%. Looking back at data for earlier years, the proportion has fallen from between 30.2% and 31% which had been the norm for the period between 2008 and 2015.

This also varies across the country, as Figure 10-10 shows, looking at data from the same survey for each of the Regional Transport Partnership areas.

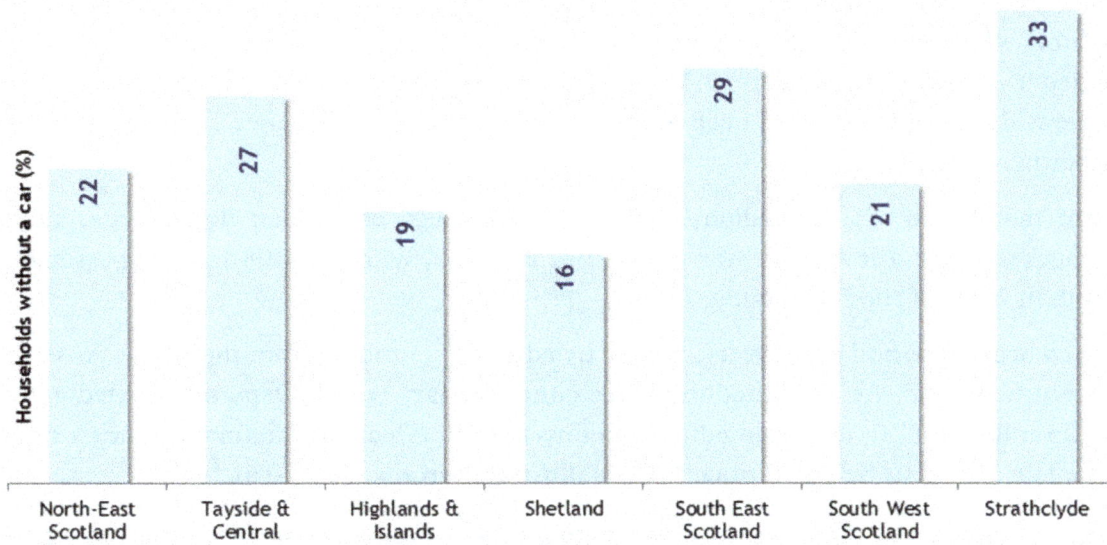

Figure 10-10: No Car Households, Scotland 2018, by RTP Area
% of survey respondents

Source: Scottish Household Survey 2018

Chapter 11: Wales

11.1   Overview

In 1982, when the publication of separate figures for Wales began, buses in the principality carried 181 million passengers, 3.3% of the total for Great Britain. The picture was part of the wider decline that had been going on throughout the UK since 1950, as discussed earlier in section 1.3 above.

From that figure of 181 million, Welsh bus patronage carried on downwards, largely unchecked, until a low point was reached in 2001/02, with just 108 million passengers, less than 2.5% of the GB total.

It then began a period of recovery, helped by additional funding from the Welsh Assembly Government, and the introduction of free concessionary travel. Demand reached a peak of 125 million in 2008/09 immediately before the full effects of the financial crisis began to be felt. The historic trend since 1982 is illustrated in Figure 11-1 below.

Figure 11-1: Historic Trends in Bus Patronage, Wales since 1982

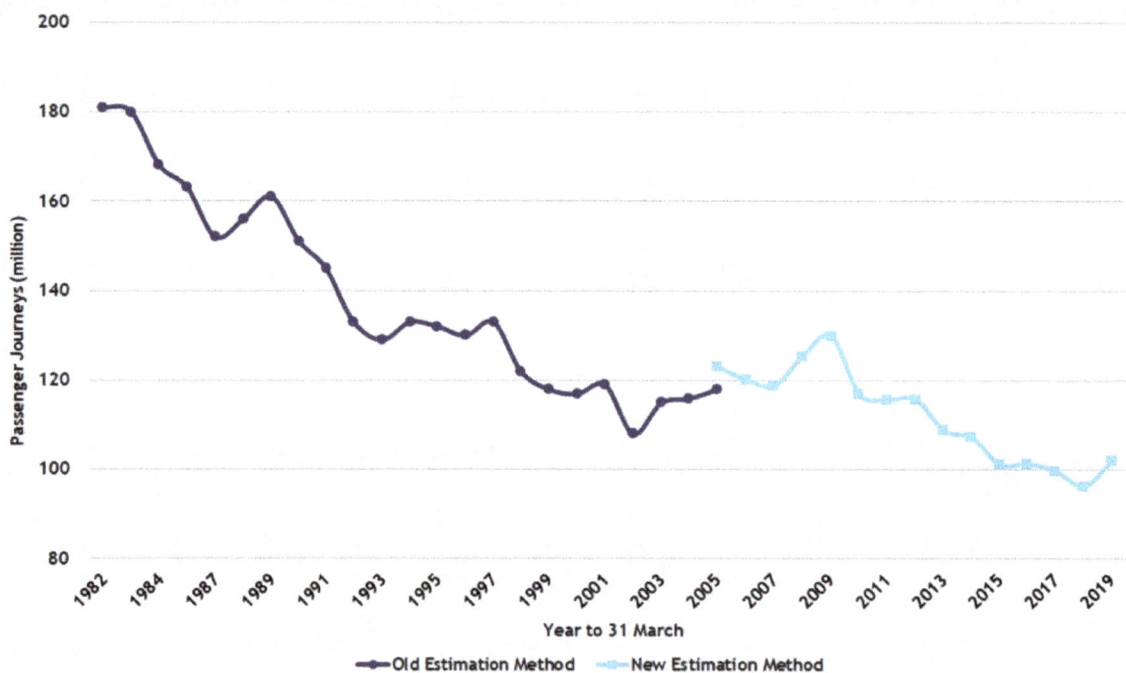

Since the recession, demand has fallen back once more, to a plateau in 2009-2012 of 115/6 million, before falling steeply to 108 and 107 million in 2012/13 and 2013/4, just 2% of the GB total. It then plunged again to reach low in 2017/18 of 96 million trips. There was a small recovery in 2018/19, bringing the total back to 102 million.

The key statistics for the period since 2005 are provided in Table 52 below. In addition to passenger journeys, there have been falls in passenger km travelled and service supply, as measured by kilometres run. Revenue is up in real terms. Since 2015, the market seems to have stabilised and the recovery in 2018/19 gave some hope for the future. Nevertheless, the worsening position has led to depot closures, service reductions and the demise of several independent operators.

Table 52: Key Traffic Statistics for Welsh Bus Services

Year to 31 March	Passenger Journeys (million)	Passenger Kilometres (Millions)	Passenger Revenue £m (2018/19 Prices)	Kilometres Run (millions)
2005	123	988	156	129
2006	120	934	167	127
2007	119	939	171	123
2008	125	970	170	124
2009	130	1,029	183	125
2010	117	937	181	124
2011	116	1,004	181	124
2012	116	1,041	185	117
2013	109	962	191	116
2014	107	948	178	112
2015	101	895	172	106
2016	101	896	171	106
2017	100	882	158	100
2018	96	885	152	99
2019	102	901	165	102
% changes				
Since 2005	-17.1%	-8.8%	5.3%	-21.3%
Last Five Years	0.7%	0.7%	-4.2%	-4.1%
Last Year	5.7%	1.8%	8.7%	2.8%

These trends are illustrated by reference to an index for each measure, with 2004/05=100, in Figure 11-2 below.

Figure 11-2: Key Indices for Buses in Wales since 2005

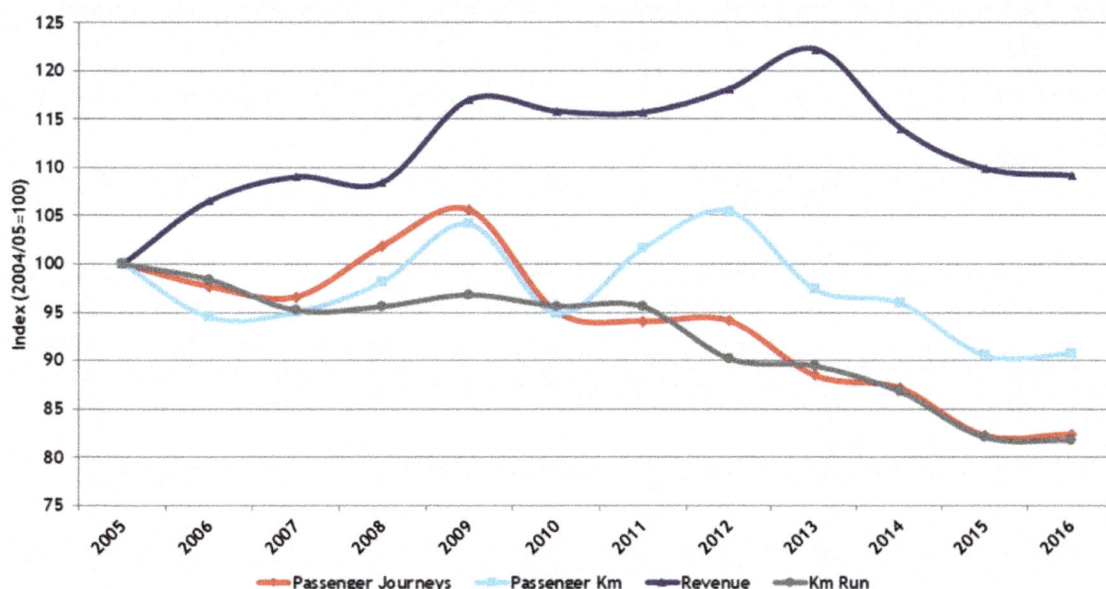

11.2   Commercial Performance

The measures which emerge from these statistics are shown in Table 53 below. Average journey length has grown over the period from at just under 8km over the 2006 to 2009 period to 8.85 for six of the last seven years. Revenue per passenger journey has also increased and was 27% higher in real terms by the end of the period. Gains in revenue per passenger kilometre have been achieved, though they are more modest. Average bus load has increased to 8.85, though falling back from the figure of 8.9+ achieved in 2011/12 and 2017/18.

Table 53: Welsh Bus Statistics – Market Analysis				
Year to 31 March	Average Journey (kilometres)	Average Fare (£, 2018/19 prices)	Yield (£, 2018/19 Prices	Average Load (Passenger Kms per Kilometre run)
2005	8.04	1.272	0.158	7.64
2006	7.77	1.388	0.178	7.33
2007	7.90	1.436	0.182	7.62
2008	7.74	1.354	0.175	7.84
2009	7.92	1.410	0.178	8.21
2010	8.02	1.550	0.193	7.57
2011	8.68	1.564	0.180	8.12
2012	9.00	1.597	0.177	8.92
2013	8.85	1.759	0.199	8.31
2014	8.85	1.665	0.188	8.44
2015	8.85	1.700	0.192	8.42
2016	8.85	1.686	0.191	8.47
2017	8.85	1.583	0.179	8.80
2018	9.19	1.574	0.171	8.94
2019	8.85	1.617	0.183	8.85
% changes				
Since 2005	10.1%	27.1%	15.5%	15.9%
Last Five Years	0.0%	-4.9%	-4.9%	5.1%
Last Year	-3.7%	2.8%	6.7%	-0.9%

11.3   Benchmarking

Unfortunately, the Welsh Assembly Government does not provide disaggregated bus patronage statistics for the different bus markets in Wales. There is therefore no published data by which to benchmark more localised performance.

11.4   Demographic and Economic Data

11.4.1   Population

The population of Wales has grown steadily over the years and the start of our review coincides with the start publication of separate bus patronage data in 1982. In that year, the principality's total stood at 2.80 million. Since then it has grown by 10.5% and stood at 3.14 million in 2018, according to the ONS Mid-Year Population Estimate. The trends over the period are illustrated in the chart at Figure 11-3 below.

Figure 11-3: Total Population in Wales, 1982-2018

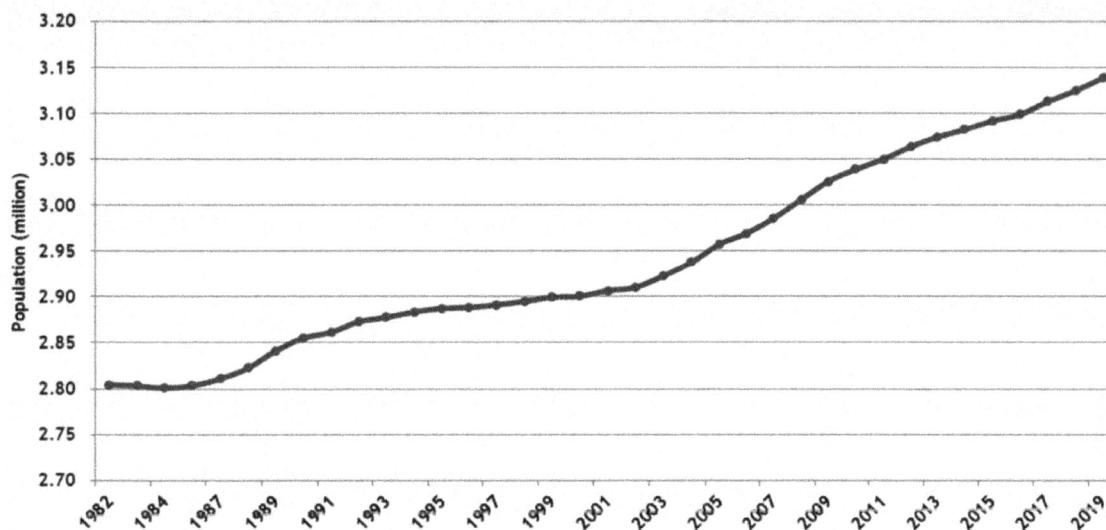

11.4.2   Employment

The pattern of employment in Wales over the last few years mirrors the rest of the UK: a sharp fall in the aftermath of the recession, followed by a strong recovery. In 2019, the total number employed was 1.5 million, an increase of 13.5% over the period. The trends are illustrated in Figure 11-4 below.

Figure 11-4: Economically Active and In Employment, Wales since 2004

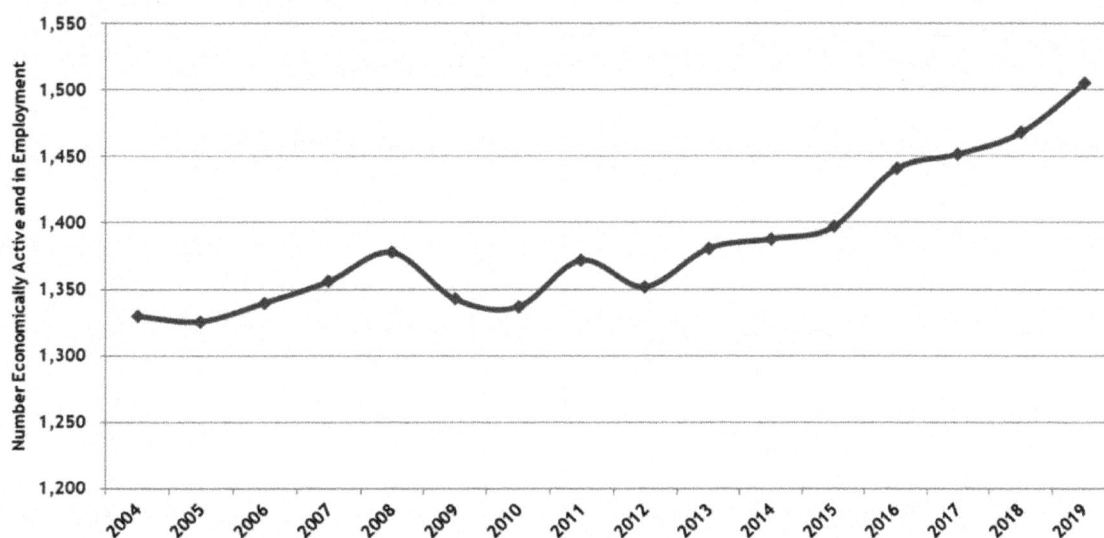

Source: Office for National Statistics, NOMIS database

11.4.3   Economic Growth

Economic growth in Wales, as measured by Gross Value Added at constant (2018) prices, clearly illustrates the effects of the recession and the gradual recovery from it. It will be seen from Figure 11-5 below that the recession took GVA down by more than 2.5% in two years in a row. Growth returned in 2010, but it was only in 2013 that the previous high achieved in 2007 was reached again and exceeded. Overall, this means that growth in the Welsh economy over the period since 2004 has been 15.2%, much lower than the 24% recorded in England.

Figure 11-5: GVA at Constant (2015 Prices), Wales since 2004

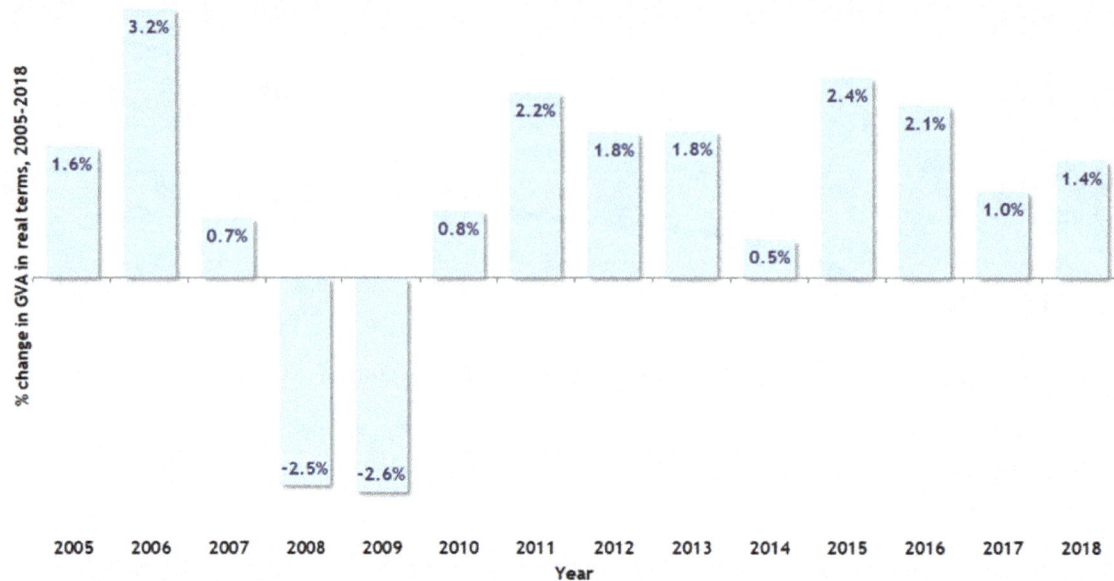

Source: Office for National Statistics. Allowance for inflation using GDP Deflator.

11.4.4   Car Ownership

Over the period since 1986, car ownership in Wales has been the third highest of any UK region/nation, at 62.2% - exceeded only by Scotland and North East England. Much of that growth took place during the 1990s and early 'noughties'. As a result, the growth over the period since 2005 has been a much more modest 5.6%, thanks in part to fall during and immediately after the recession. The details of the trends are contained in Table 54 below.

Table 54: Car Ownership per 1,000 people, Wales since 1986

Year	Cars
1986	299
1994	349
2001	381
2004	460
2005	470
2006	472
2007	477
2008	478
2009	476
2010	477
2011	469
2012	471
2013	474
2014	480
2015	485
2016	483
2017	493
2018	497
% Changes	
Since 1986	66.1%
Since 2001	30.5%
Since 2005	5.6%

Source: Vehicle Licensing Statistics, DfT, and Mid-year Population Estimates, ONS

Chapter 12: The London Market

12.1   Introduction

For a whole variety of reasons, the market for bus in London is completely different from the rest of the country. The regulatory system, the uniquely dense population and the concentration of employment are amongst the reasons for this.

The demographic analysis in Chapter 4 focused primarily on England outside London. This chapter looks at the same factors for London. Travel by bus in London features in the National Travel Survey as a separate mode from travel by bus outside the capital and, as will be seen, there are marked differences in travel patterns and behaviour.

In the review of NTS data on bus use in London, it is important to remember that the survey covers people in the whole of England, so that the responses will include large numbers of people who never or rarely visit the capital. There will be many others – particularly those who live in the Home Counties but work in London – who are in the city as often as many residents. Thus, the overall trip rates appear to be much lower, because the figures represent travel by bus in London by all people in England, not just Londoners.

12.2   Historical Overview

As was discussed in section 1.4 above, the London market saw very significant growth in patronage over the years between 1994 and 2014. However, this was a marked reversal of previous trends: prior to 1994, street-running public transport had seen a decline in numbers which dated back to the late 1940s.

Thus, in 1948, the London Transport Executive's red bus, trolleybus and tram services carried 3,641 million passenger journeys[7]. By 1962, this had declined by over 39% to 2,215 million. After this, numbers carried on falling – reaching a low point of 1,041 million in 1981, immediately before the introduction of the GLC's "Fares Fair" policy. This sparked a recovery in passenger numbers that survived the reversal of the cheap fares policy in the spring of 1982 after court action. This was the point at which the Travelcard was first introduced.

The recovery continued until the end of "Lawson boom" of the 1980s, reaching a peak of 1,211 million in 1988/89. Thereafter, the recession dampened demand and there were five successive years of further decline, down to 1,117 million journeys in 1993/94.

In that year, the decline stopped, and began to be reversed. By 2014/15, the total had reached 2,364 million passenger journeys, more than double the 1993/94 figure. The last time that number of people travelled by bus was in 1959. That proved to be peak for the time being, and numbers have fallen every year since, the total decline being 7%. The long-term trends are illustrated in the graph at Figure 12-1 below.

---

[7] All the figures for patronage prior to 1970 are taken from A History of London Transport Volume 2: The Twentieth Century to 1970. Barker and Robbins, 1974.

Figure 12-1: Bus Patronage in London since 1950

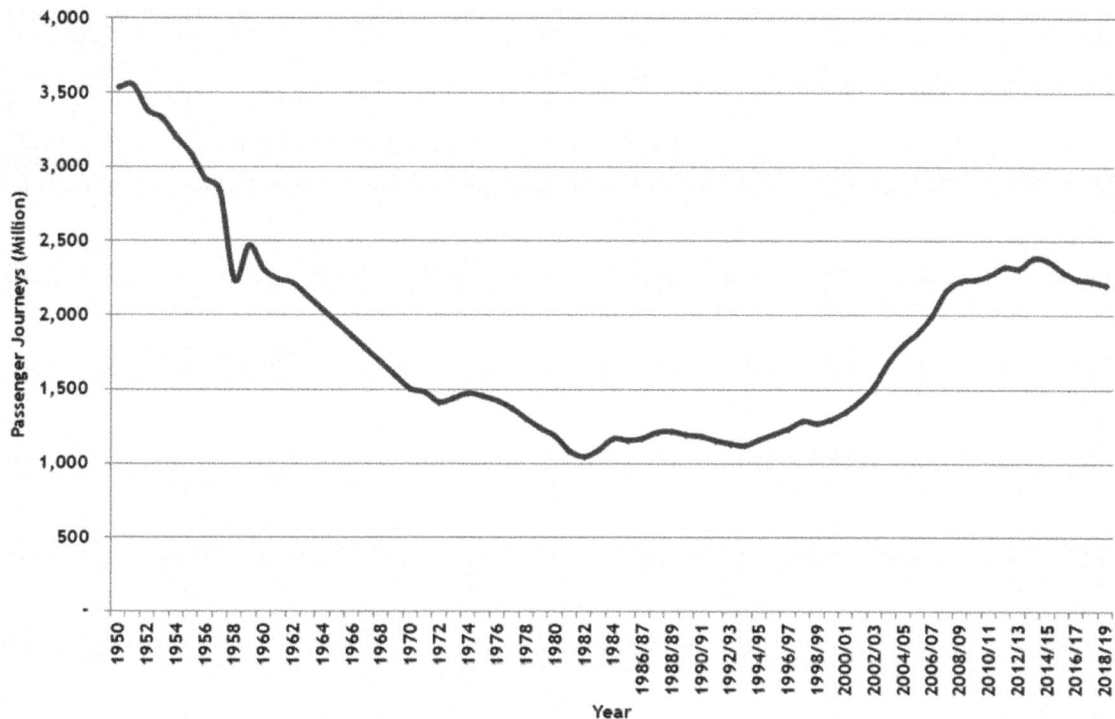

Includes tram services (1950-52) and trolleybus services (1950-62)

12.3   Demographic Influences

The subjects covered are:

•   Growth in population and increases in density

•   Age and Gender

•   Household Income

•   Socio-Economic Classification

12.3.2   Population and Density

Total Population

A long-term decline in the number of people living in the capital began in the post war years, as policies of slum clearance and rebuilding after the blitz saw population densities fall and large numbers of people moved out to new towns and overspill estates.

From a peak of 8.61m in 1939, the population had fallen to 7.98m by 1961, and continued to fall until 1988, when it reached a low point of 6.73m – more than 21% below the immediate pre-war figure.

The capital's shrinkage then stopped, and the population began to grow again – doing so in every year except one (2000) since. The number of people living in Greater London went above seven million for the first time in 20 years in 1997 and exceeded 7.5 million in 2005 – the highest since 1971. The eight million barrier was crossed in 2011 for the first time since the 1950s, and that historic 1939 record was beaten in 2015. The latest estimate for 2018 saw the total at 8.91 million. The pattern of decline and recovery is illustrated in Figure 12-2 below.

This means that, since the low point in bus patronage seen in 1993/94, the population of Greater London has risen by more than a quarter.

Figure 12-2: Long Term Trends in London Population Levels

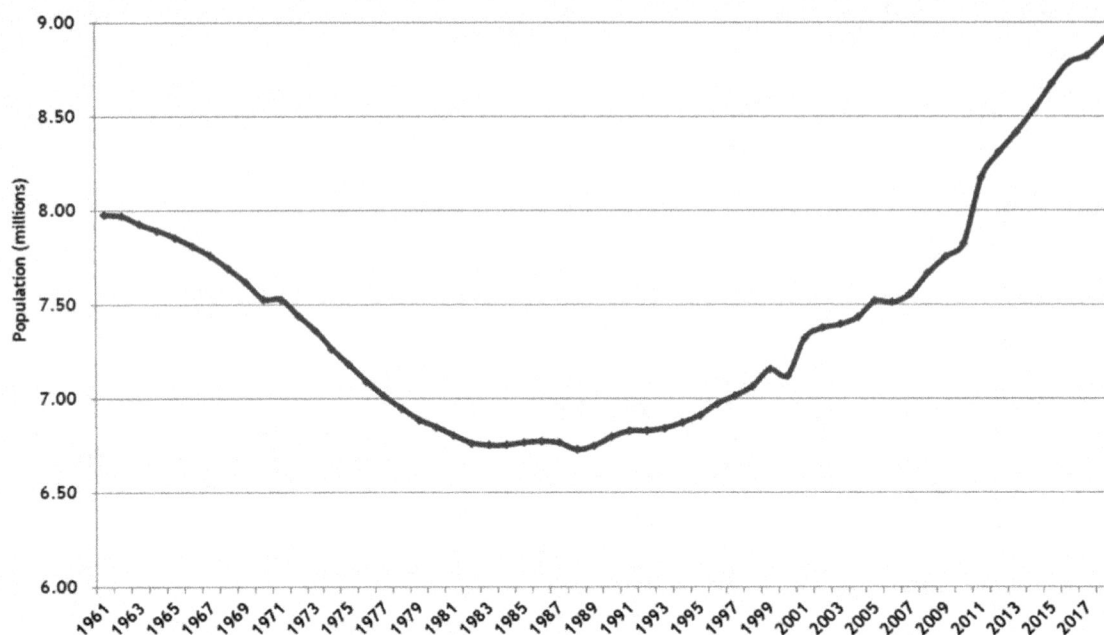

Source: Transport for London and Office for National Statistics

Population Density

As noted earlier in Chapter 4, London is much more densely populated that the rest of the country, and the population growth discussed above has only increased this. We also saw that population density is a crucial factor in determining the market potential of a bus network.

Across the whole of the GLA area, density has increased by 1,000 people per square kilometre since 2001, going from 4,658 to 5,667 in 2018. For Inner London, the figure has gone from 8,964 to 11,286.

12.3.3 Bus Use by Age and Gender

Trip Rates

There is a large variation in the use made of buses by different age groups and genders. This is measured annually as part of the Department for Transport's National Travel Survey. The figures for the most recent year are shown in Figure 12-3 below.

As will be seen, women of most ages make more use of buses than men – though the difference can be exaggerated: it is not true to say that the bus product is used *overwhelmingly* by women.

In London, males between 17 and 20 appear to make more use of buses than any other age or gender group. Other high trip rates are recorded amongst women between 21 and 29 and from 30 to 39. Thereafter, as in the rest of the country, trip rates fall quite sharply for the 40-49 and 50-59 age groups. Men in the 60-69 and 70+ age group tend not to increase their trip-making very much, despite the relatively more generous concession scheme on offer in the capital.

In terms of age group, the biggest users of bus services are people between the ages of 17 and 20, followed by those between 30 and 39, whilst the third most important group is the 21-29 cohort. The age group with the lowest volume of trips in 2018 was people in the 50-59 band, particularly men, who made half as many trips as their thirtysomething counterparts.

Figure 12-3: Bus Trips by Age and Gender, Bus in London, 2018

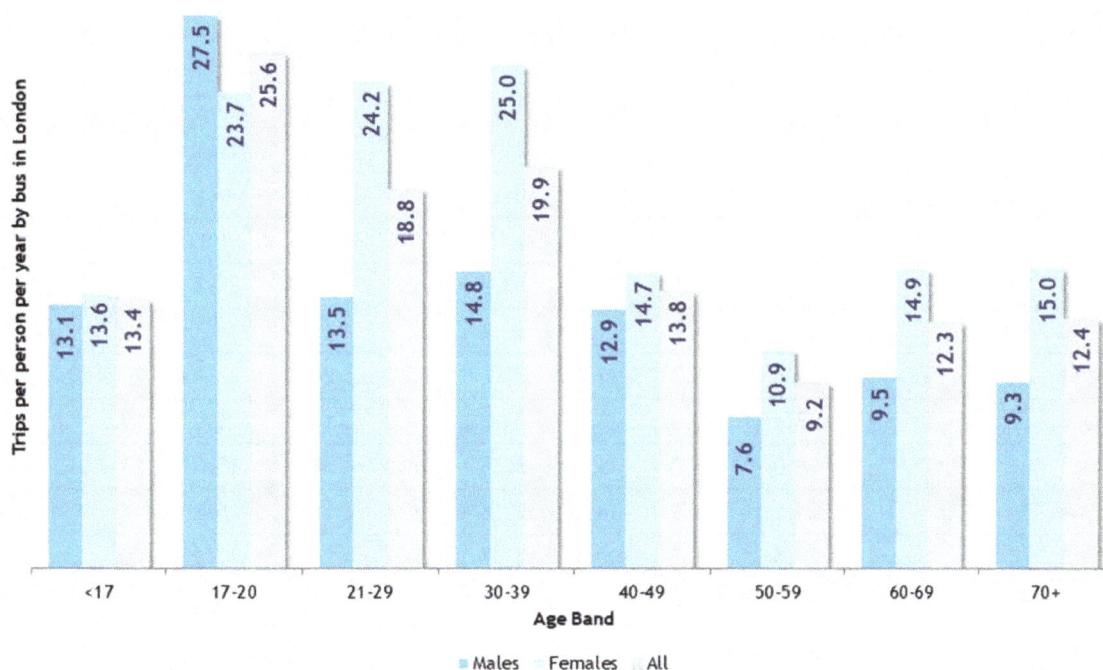

Source: National Travel Survey 2018, DfT

Trends over Time

The changes in trip rates over time are recorded in Table 55 below. It is apparent that there have been some falls in trip rates over time. The table shows the results of each survey with a median figure for the 2009-2017 survey and a comparison with the 2018 results. Overall rates fell by a quarter using this approach, with some falling by as much as one third. The falls seem to be particularly significant since 2013 and have occurred alongside the fall in patronage since 2014 already discussed.

The implication seems to be that social changes in the capital, and amongst visitors/external commuters have certainly impacted on the market for bus travel Across both genders, falls amongst children under 17, the two "middle age" decades of 40-49 and 50-59 have fallen by over 40%.

As with the rest of the country, the lowest trip rates are amongst the 50-59 age group of both genders, whilst the highest rates are amongst the 21-29 cohort. Trip rates rise for people aged 60+, but not by as much as the existing of Freedom Passes might lead one to expect.

Table 55: Bus Trip Rates by Age & Gender, Bus in London

Trips per person per year	All ages	<17	17-20	21-29	30-39	40-49	50-59	60-69	70+
Males									
2009	21.7	18.9	32.4	36.6	22.5	20.4	16.4	15.6	18.7
2010	24.7	20.0	44.2	44.5	28.3	21.8	16.0	16.7	19.5
2011	20.7	16.3	25.4	34.1	27.6	17.5	14.2	16.5	19.1
2012	18.6	18.5	22.6	23.2	23.4	13.9	13.6	18.7	18.0
2013	21.3	25.5	40.1	24.9	21.4	18.8	14.1	12.9	21.6
2014	18.9	20.5	33.8	23.7	22.4	14.9	16.4	12.4	15.1
2015	20.1	20.4	31.0	24.5	19.8	19.7	17.2	18.3	16.6
2016	16.1	17.4	24.5	17.6	20.0	15.5	12.4	12.9	12.4
2017	17.5	16.5	24.0	22.9	21.6	18.6	12.5	18.4	10.2
2018	15.2	13.4	25.6	18.8	19.9	13.8	9.2	12.3	16.2
Median 2009-17	20.4	18.9	31.0	24.5	21.6	17.5	14.1	15.6	18.0
% chg. from median	-25.6%	-29.4%	-17.4%	-23.3%	-7.8%	-21.3%	-34.4%	-21.3%	-9.9%
Females									
2009	24.2	21.1	36.0	33.8	27.8	23.8	20.2	19.0	20.4
2010	25.6	21.3	41.3	39.2	32.7	25.6	15.3	17.7	22.5
2011	22.4	16.7	32.2	32.2	28.9	24.7	13.4	20.7	18.8
2012	19.8	20.5	28.5	22.2	24.8	14.0	15.5	19.6	19.3
2013	24.0	26.0	45.6	30.5	26.1	26.8	15.6	13.3	19.3
2014	21.6	21.9	38.2	22.8	30.4	18.2	20.8	14.4	15.7
2015	22.5	22.5	30.3	27.7	24.1	25.5	17.1	20.5	17.5
2016	18.7	19.6	27.9	20.3	25.4	16.0	14.8	16.1	15.0
2017	18.0	18.6	21.4	24.9	23.0	18.1	14.1	12.5	13.1
2018	16.9	13.6	23.7	24.2	25.0	14.7	10.9	14.9	15.3
Median 2009-17	22.5	20.9	32.2	27.7	26.1	18.2	15.4	17.7	17.5
% chg. from median	-24.7%	-34.8%	-26.5%	-12.7%	-4.2%	-19.2%	-29.7%	-15.8%	-12.3%
All Persons									
2009	21.7	18.9	32.4	36.6	22.5	20.4	16.4	15.6	18.7
2010	24.7	20.0	44.2	44.5	28.3	21.8	16.0	16.7	19.5
2011	20.7	16.3	25.4	34.1	27.6	17.5	14.2	16.5	19.1
2012	18.6	18.5	22.6	23.2	23.4	13.9	13.6	18.7	18.0
2013	21.3	25.5	40.1	24.9	21.4	18.8	14.1	12.9	21.6
2014	18.9	20.5	33.8	23.7	22.4	14.9	16.4	12.4	15.1
2015	20.1	20.4	31.0	24.5	19.8	19.7	17.2	18.3	16.6
2016	16.1	17.4	24.5	17.6	20.0	15.5	12.4	12.9	12.4
2017	17.5	16.5	24.0	22.9	21.6	18.6	12.5	18.4	10.2
2018	15.2	13.4	25.6	18.8	19.9	13.8	9.2	12.3	16.2
Median 2009-17	20.4	18.9	31.0	24.5	21.6	17.5	14.1	15.6	18.0
% chg. from median	-25.6%	-29.4%	-17.4%	-23.3%	-7.8%	-21.3%	-34.4%	-21.3%	-9.9%

Source: National Travel Survey 2018, Department for Transport

Market Share

By multiplying the trip rates discussed above by the proportions of the population in each age band and gender, it is possible to assess the importance of each segment of the population to the current levels of bus use. This can then enable operators and authorities to target improvements and promotional activity to those segments.

Looking first at age (both genders) in Figure 12-4, we can see the importance of the under 17 market to current patronage levels - accounting for almost 18%. The next most important group are the 30-39 cohort, accounting for 16.4% of trips. Next come the over 70s, who account for 14.9% of trips. They are followed by the 21-29 group on 12.5%.

The spread amongst the age groups indicates that bus travel is overwhelmingly a younger person's product, with 40% of trips being made by the under 30s. This dwarfs the 22% undertaken by the over 60s. But it is clear from both the trip rates and the market share proportions that the key task – and key opportunity for the industry – is to win back those between 40 and 59.

Figure 12-4: Bus Journeys by Age, London, 2018

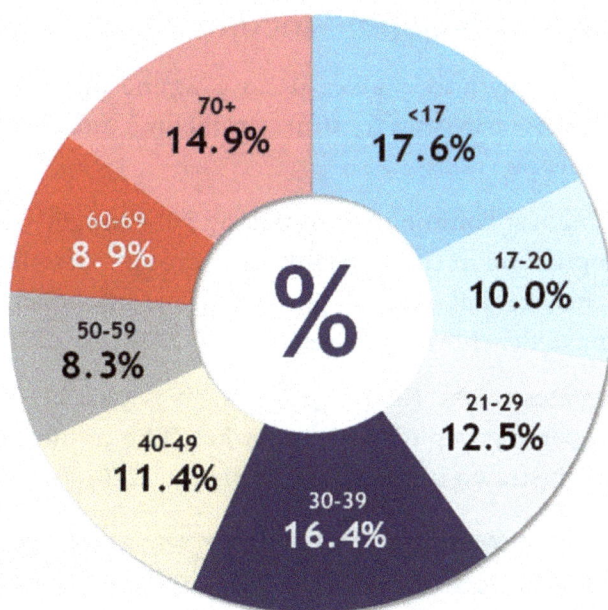

Source: PTIS, based on DfT National Travel Survey and ONS Mid-Year Population Estimates 2018

The relative importance of the gender balance can be seen in Figure 12-5 below. This takes the same estimates but looks at rather more consolidated age bands. This shows that in each of the principal age markets – 'young' (17-30), "middle" (31-59) and "elderly" (60+) – women clearly predominate with this being particularly strong in the 31-59 group.

121

Figure 12-5: Bus Journeys by Age & Gender, London, 2018

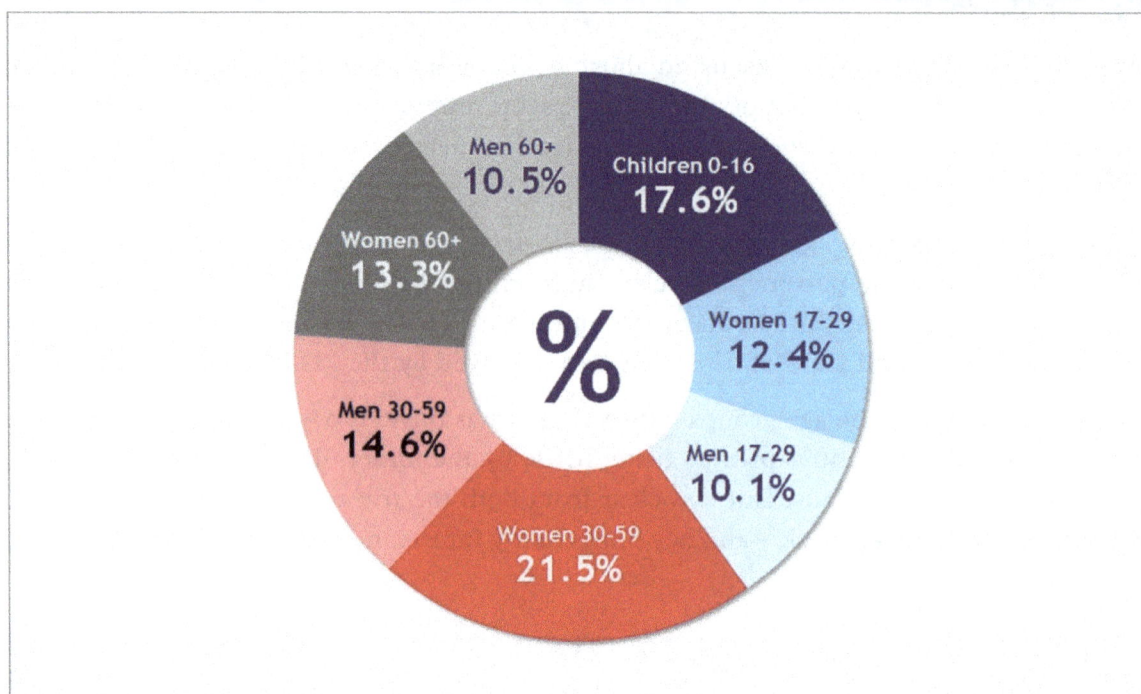

12.4   Bus Use by Household Income

The National Travel Survey also enables us to examine bus use by household income levels, as the statistics show trip rates for different income quintiles. The analysis for 2018 is shown in Table 56 below.

As can be seen, the highest volume of trip-making is, as might be expected, undertaken by the 20% of the population in the lowest income group – people in this income bracket make half as many trips again as those in the highest income bracket. However, the lower income group also make the shorter journeys.

By applying the trip rates to the total population, it is possible to estimate the relative importance of each income group to the bus market. The results are contained in the last line of Table 56, and illustrated in the graph at Figure 12-6.

Table 56: Bus Use in London, by Income Quintile

Annual figures	Lowest real income level	Second level	Third level	Fourth level	Highest real income level	All income levels
Trips per person	22	14	13	13	14	15
Kilometres per year	92	55	51	50	60	62
Average Journey (km)	4.3	3.5	4.5	4.4	5.0	4.3
Implied % of trips	30.0%	17.8%	16.4%	16.3%	19.5%	100.0%

Source: National Travel Survey 2018 Sheet NTS0705, Department for Transport (Rows 1-3). PTIS Analysis (Row 4)

Figure 12-6: Bus Journeys in London, by Income Quintile

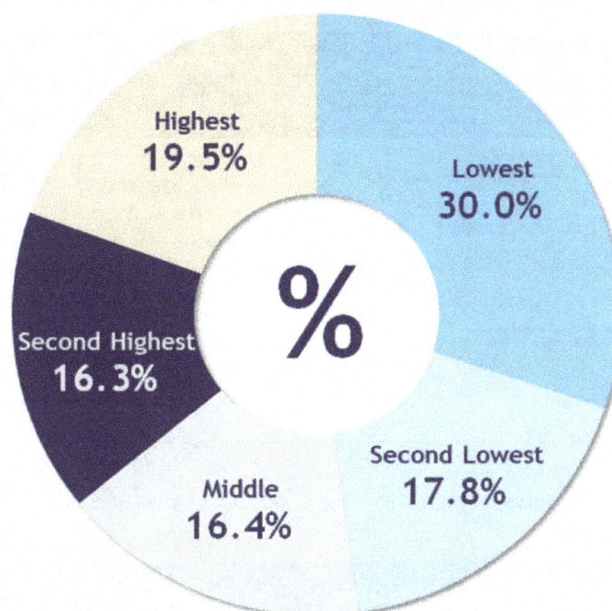

Source: PTIS estimates based on data in National Travel Survey 2018, DfT

Thus, it can be said that, whilst the largest proportion of bus users is in the lowest income group, other groups still comprise a significant proportion of patronage – it is particularly notable is that the second largest volume of trips in London is amongst highest income group. It is not true to say that buses in London are purely for the poorer groups in our society.

12.5   National Statistics Socio-Economic Classification

NTS data also provides a picture of the nature and extent of bus use by socio-economic classification, using the three-class version. This provides a high-level portrait, using the following classes:

• Higher managerial, administrative and professional occupations

• Intermediate occupations

• Routine and manual occupations

Other members of the population are included as either "Never worked and long-term unemployed" or "Unclassified". These are mainly full-time students.

The ONS annual population survey provides us with a percentage breakdown of the population by these groups, and the 2018 NTS provides a trip rate for each group. It is thus possible to estimate the relative importance of each socio-economic group to the market for bus travel.

The results are shown in Table 57 and the market shares are illustrated in the graph at Figure 12-7.

Table 57: Bus Use in London, by Socio-Economic Group
People Aged 16-64

Item	Managerial and professional occupations	Intermediate occupations	Routine and manual occupations	Unemployed and economically inactive	Unclassified (mainly students)
% population breakdown	36.6%	27.5%	12.6%	17.7%	5.6%
2018 Trip Rate – journeys per person per year	11.9	12.2	16.3	40.7	19.7
% of trips	24.1%	18.5%	11.4%	39.8%	6.1%

Source: ONS Annual Population Surveys (via NOMIS) Line 1. National Travel Survey Line 2. PTIS analysis Line 3.

Figure 12-7: Patronage in London by Socio-Economic Group
People Aged 16-64

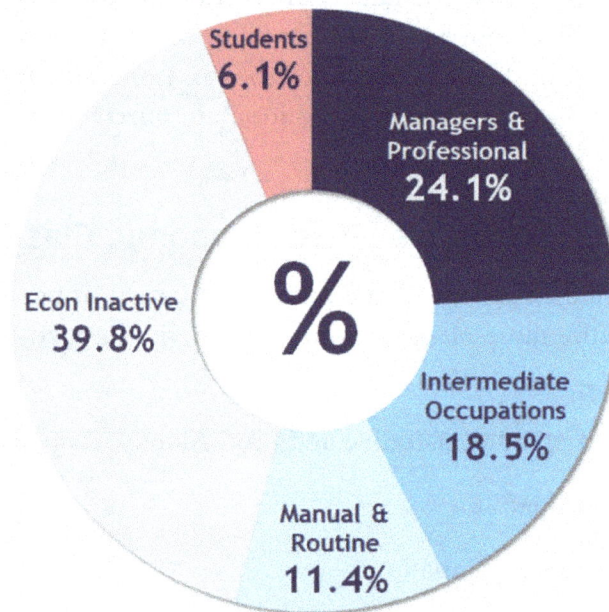

The analysis shows that, amongst those aged 16-64, patronage is more evenly split between different occupations than might be expected. Apart from students, people who are unemployed and otherwise economically inactive have the highest trip rate at 40.7, and our estimates suggest that they account for 39.8% of patronage. Those in manual and routine occupations have the next highest trip rate (16.3). However, there are relatively few of them these days, so that they only account for 11.4% of trips.

Those in the professional and intermediate level occupations use the bus the least frequently. However, they do now represent a very substantial proportion of the working population, so continue to be very important to the bus industry, accounting for 24.1% of trips between them. Students seem less important to the London market than in the rest of the country: they account for 5.6% of the population, and account for 6.1% of all trips.

12.6   Key Statistics

Figures for the period since 2004/05 are shown in Table 58 below. Despite recent falls, passenger journeys are still over 21% up, with the number of passenger kilometres up by 22.6%. Revenue has grown by 27% in real terms against real-term fare increases of 25.5%. An index of these three measures is illustrated in Figure 12-8. At the same time, service supply as measured by kilometres run has been increased by 1.4%.

However, the picture over the last five years is much less favourable: patronage is 7% down in journeys and 4.1% in passenger kilometres. Revenue is 11.6% down in real terms and service supply reduced by 1.8%. This seems to be largely attributable to increased congestion, especially in the central area.

Table 58: Key Traffic Statistics for London Bus Services

Year to 31 March	Passenger Journeys (million)	Passenger Kilometres (Millions)	Passenger Revenue £m (2018/19 Prices)	Kilometres Run
2005	1,802	7,782	1,137	470
2006	1,881	8,420	1,205	461
2007	1,993	8,727	1,250	465
2008	2,160	9,446	1,273	465
2009	2,228	9,432	1,259	474
2010	2,238	9,365	1,310	479
2011	2,269	9,369	1,433	481
2012	2,324	9,560	1,497	485
2013	2,315	9,697	1,554	486
2014	2,384	10,002	1,620	487
2015	2,364	9,949	1,632	485
2016	2,293	9,675	1,607	488
2017	2,240	9,454	1,514	490
2018	2,226	9,788	1,465	486
2019	2,198	9,540	1,442	476
% changes				
Since 2005	21.9%	22.6%	26.9%	1.4%
Last Five Years	-7.0%	-4.1%	-11.6%	-1.8%
Last Year	-1.3%	-2.5%	-1.5%	-1.9%

Figure 12-8: Index of London Bus Traffic Statistics

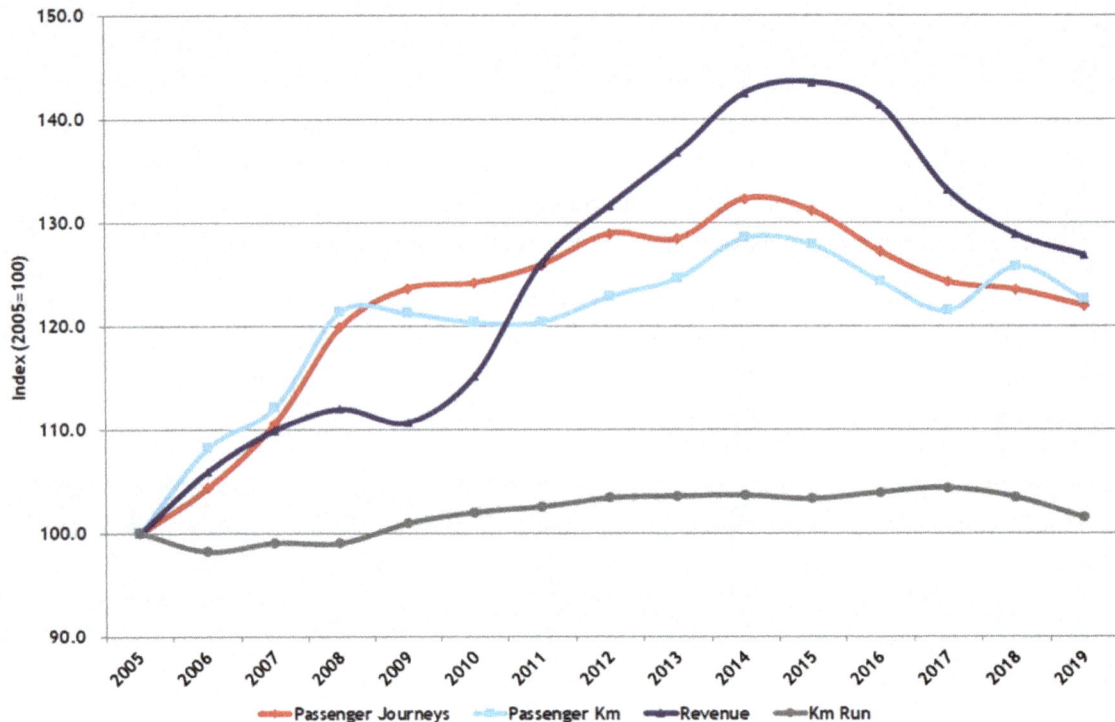

12.7   Commercial Performance

Table 59 below shows the analysis of market trends derived from the traffic statistics.

It will be seen that the growth has led to an improvement in the average load on each vehicle of some 21%, taking the number up to 20 compared with 16.6 in 2005 (and as low as 11.0 in 1994/95).

In real terms (2018/19 prices), the earnings per passenger have risen by 4.0% over the last fifteen years, and now stand at 65.6p. This is well ahead of the low point of 56.5p in 2008/09. This year marked the last year in which fares had been influenced by Ken Livingstone as London Mayor. Boris Johnson was elected in April 2008 and his first decision on fares was implemented in January 2009. Revenue per passenger journey reached a peak of 70.1p in 2015/16, the last year of Johnson's mayoralty. Since then, the fares freeze promised by the incoming mayor Sadiq Khan has resulted in a 5% real reduction.

Real revenue per passenger kilometre is up over the period. At 15.1p in 2018/19, this is 3.5% higher than in 2004/05. Again, the influence of fares policies is clear, with yields falling as low as 13.3p under Mayor Livingstone in 2008/09. Under Boris Johnson, they rose by 24.5% to a peak of 16.6p in 2015/16. Since then, Mayor Khan's freeze has seen the figure fall back by 9%.

Looking at the trends for real revenue per kilometre operated, this has grown substantially as patronage has risen, and is 20.2% higher than 2004/05, despite recent slippage. It reached a peak in real terms in 2013/14, when the figure reached £4.90. However, it has since fallen back by 5.8% to fell back to £4.61.

Londoners have been tending to use the bus for slightly longer journeys again recently. The average journey length started the period at 4.32km, then grew to a maximum of 4.48km just before the recession. During the downturn, journeys began to get shorter, falling to a low point of 4.11 km in 2012. Since then, journey lengths have started to grow again, and reached over 4.3 km once more for the last two years. This remains substantially ahead of the long-term average of 3.4 km recorded throughout the 1990s, with the change coinciding with the introduction of the congestion charge in 2003.

Table 59: London Bus Statistics – Market Analysis					
Year to 31 March	Average Journey (Kilometres)	Average Fare (£, 2018/19 prices)	Yield (£, 2018/19 prices)	Average Load (Passenger Journeys per Passenger Km)	Revenue per Km (£, 2018/19 prices)
2005	4.32	0.631	0.146	16.57	3.84
2006	4.48	0.640	0.143	18.26	4.08
2007	4.38	0.627	0.143	18.77	4.29
2008	4.37	0.589	0.135	20.32	4.65
2009	4.23	0.565	0.133	19.90	4.70
2010	4.18	0.585	0.140	19.56	4.67
2011	4.13	0.632	0.153	19.47	4.71
2012	4.11	0.644	0.157	19.70	4.79
2013	4.19	0.672	0.160	19.95	4.76
2014	4.20	0.680	0.162	20.55	4.90
2015	4.21	0.691	0.164	20.51	4.87
2016	4.22	0.701	0.166	19.83	4.70
2017	4.22	0.676	0.160	19.29	4.57
2018	4.40	0.658	0.150	20.15	4.58
2019	4.34	0.656	0.151	20.02	4.61
% changes					
Since 2005	0.5%	4.0%	3.5%	20.8%	20.2%
Last Five Years	3.1%	-4.9%	-7.8%	-2.4%	-5.4%
Last Year	-1.3%	-0.3%	1.0%	-0.6%	0.6%

12.8   Economic Data

12.8.1   Employment

In the period since March 1994, the number of people employed in Greater London has risen by 55%, from 2.953m to 4.676m in April 2019. Growth over the period has been virtually continuous, with small setbacks in 2003, and then again in two of the years during the recession, 2010 and 2012. The changes are illustrated in Figure 12-9 below.

The economic activity level amongst the population has risen too, from 64.8% in 1994 to 68.7% in April 2011 – having reached just over 70% immediately prior to the recession. By 2016 it had reached 78.2%. At the beginning of 2019, it stood at 79.6%.

Figure 12-9: Changes in London Employment since April 1994

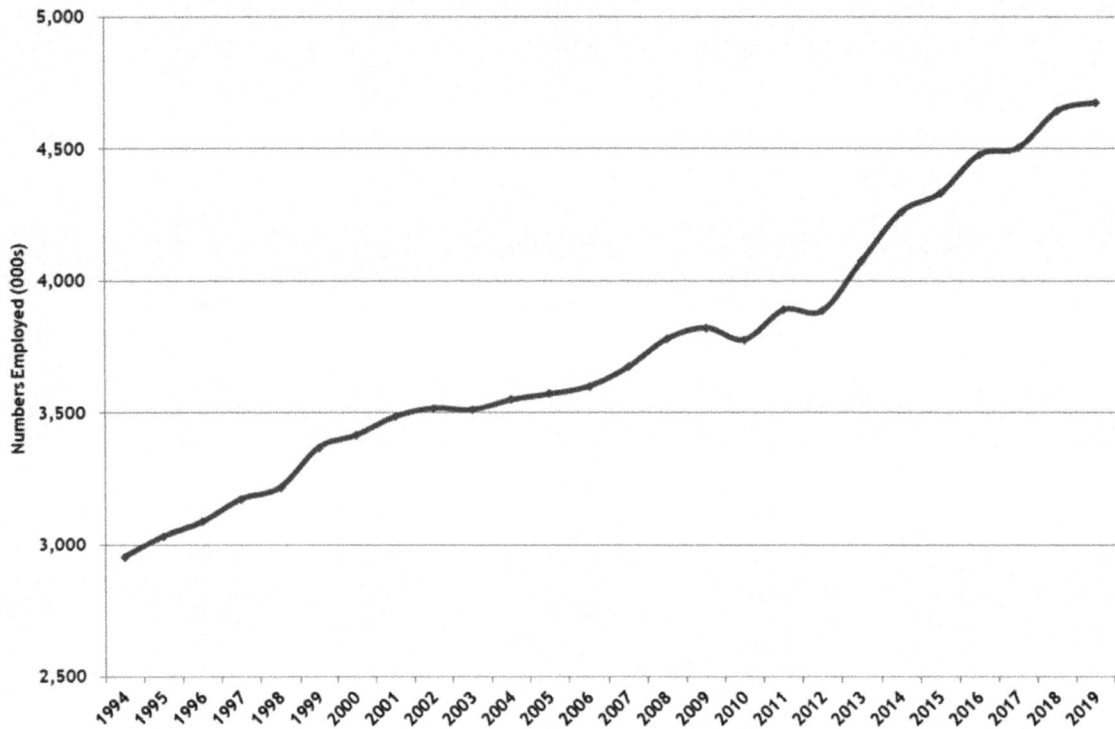

Source: NOMIS Analysis of Labour Force Survey, ONS

12.8.2 Economic Growth

The London economy has grown rapidly over the same period, with only one year (2009) showing a negative number. After allowing for inflation using the GDP Deflator, the analysis suggests that the total growth of just over 74% has been achieved over the period since 1998. The figures are illustrated by the graph at Figure 12-10 below.

Figure 12-10: Trends in Gross Value Added for London since 1998

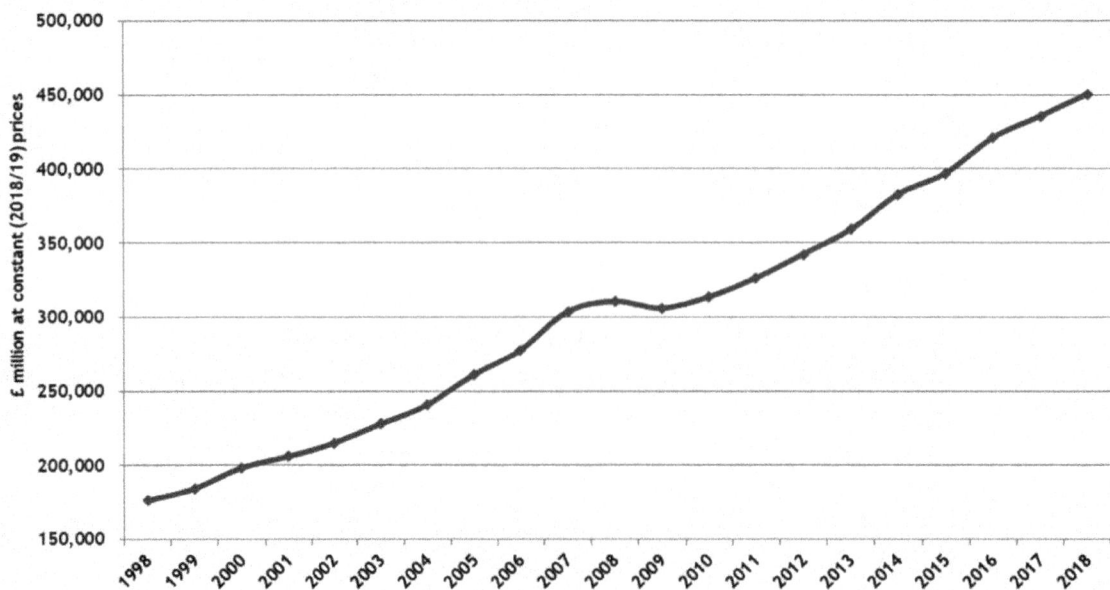

Source: Office for National Statistics

12.9   Car Ownership

In 1986, 324 cars were owned for every 1,000 London residents, which placed the region fourth highest in the country. Only South East England (446), South West England (365) and East of England (377) had higher rates. By the turn of the century, though the rate in London had grown, rapid growth in other parts of the country meant that the capital now had the lowest car ownership rate in the country – a position which it has retained ever since.

The rate in London did continue to grow slowly, despite the introduction of congestion charging, peaking at 341 in 2007 and 2008. By 2009, however, the onset of the recession and higher oil prices meant that ownership levels fell back to 330 per 1,000 – below the starting figure. Ownership levels were roughly stable at around 304 between 2013 and 2017 but fell again in 2017 to below 300 for the first time for decades. Meanwhile, as we saw in Figure 6-4 above, rates in other parts of the country have carried on growing.

Table 60: Car Ownership Rates in London

Year	Cars Owned per 1,000 people
1986	324
1994	330
2001	321
2004	340
2005	341
2006	341
2007	341
2008	338
2009	330
2010	327
2011	311
2012	305
2013	303
2014	304
2015	304
2016	304
2017	302
2018	299

Source: PTIS analysis of DfT vehicle registration statistics and ONS population figures.

12.10   Conclusions

This review of the London market has illustrated how, in terms of population, employment and economic growth, the market has been moving in favour of the bus in London for a couple of decades. Until 2014, these favourable movements were reflected in year on year patronage growth. However, since then, and despite falls in real-term fares following Mayor Khan's fares freeze, the market has turned. Anecdotally, this seems to be mainly attributable to slowing bus speeds as a result of growing congestion.

The changes that have occurred can be seen in Table 61 below, which shows the TfL data for average bus speeds in different parts of London. As can be seen, whilst average speeds have fallen in the GLA area as a whole by 2.7%, the some inner London boroughs have seen much greater falls – most notably the City of London on almost 11%, and three with falls of more than seven per cent – Lambeth, Lewisham and Southwark. Islington has seen a 6.3% fall and Tower Hamlets 5.3%.

Table 61: Observed Bus Speeds (mph) in London since 2013/14
Median across the year for each borough

Borough/Year to 31 March	2020	2019	2018	2017	2016	2015	2014	% change
Camden	6.98	6.85	7.03	6.94	7.06	7.26	7.26	-3.8%
City of London	5.94	5.86	5.94	5.52	5.69	6.37	6.65	-10.7%
City of Westminster	7.18	7.11	7.17	6.96	7.00	7.22	7.37	-2.6%
Greenwich	8.40	8.44	8.66	8.51	8.50	8.69	8.84	-5.0%
Hammersmith & Fulham	8.24	8.41	8.24	8.33	8.35	8.40	8.52	-3.3%
Islington	7.77	8.00	7.97	7.88	7.93	8.24	8.30	-6.3%
Kensington & Chelsea	7.50	7.56	7.35	7.51	7.52	7.60	7.64	-1.9%
Lambeth	8.08	8.09	8.18	8.26	8.40	8.59	8.74	-7.6%
Lewisham	9.00	9.01	9.16	9.09	9.18	9.54	9.71	-7.4%
Southwark	8.10	8.06	8.08	7.89	7.87	8.51	8.73	-7.2%
Tower Hamlets	8.69	8.70	8.70	8.69	8.44	8.98	9.17	-5.3%
Wandsworth	8.61	8.79	8.79	8.71	8.77	8.92	9.01	-4.4%
ALL GLA	9.29	9.25	9.21	9.23	9.30	9.42	9.55	-2.7%

Source: Transport for London

Chapter 13: Putting the Jigsaw Together

13.1   Overview

At the start of this report, it was suggested that understanding bus demand was akin to putting a jigsaw together. To extend the metaphor, subsequent chapters have discussed the various pieces, identifying the evidence to support each one.

This chapter aims to complete the puzzle, estimating how changes to each component have affected the overall picture. In order to do so, a series of percentage changes is derived from the data set out in previous chapters, and the expected change in demand for bus services is estimated by reference to a series of elasticities of demand. These have been taken so far as possible from *The Demand for Public Transport: A Practical Guide*, published by TRL as report no 593 in 2004.

This is very much the approach which the present author used with some success when working as Project Director with The TAS Partnership (TAS) on a National Bus Model for the Department for Transport between 2002 and 2005. A similar approach has been taken with a range of other clients since by both TAS and this company since the businesses went their separate ways in 2016.

The aim of this chapter is to focus primarily on the results of this exercise, rather than providing a detailed account of the modelling work. However, it may be useful to set out the demand elasticities used, and these are contained in Table 62 below.

Table 62: Elasticities of Demand for Different Components

Component of Demand	Paying Customers	Concessionary Pass Holders
Employment	1.000	-
Economic Growth (Real Changes in GVA)	0.150	0.150
Population	1.000	1.000
Car Ownership (Work Trips)	-1.000	-
Car Ownership (Non-Work Trips)	-0.400	-0.300
Fares (short term, up to 1 year)	-0.420†	-0.230
Fares (medium term, after 1 year)	-0.140	-0.030
Work trip as proportion of whole (NTS 2018)	22.9%	-
Other trips as proportion of whole (NTS 2018)	77.1%	-
Service supply (km run) - short run	0.480	0.480
Service supply (km run) - long run	0.207	0.207
Pensioner Incomes	-	0.150
Car Fuel Prices (Short Run)	0.330	0.013
Car Fuel Prices (Long run - after 1 year)	0.100	-
Changes in Bus Journey Times	0.250	-

† = 0.630 in London

It must be emphasised that these are assumptions, based on previous research, and there may be those who disagree with some or all of them. However, they have been sourced as closely as possible from other published and accredited research work and have not been chosen arbitrarily.

Since bus is primarily a local product, and data on such matters as fares and service changes is not easily available in sufficient detail, the results of this work will not necessarily be applicable to all local networks within a given area or territory type. This is commented on in the text for each area. Thus, for example, fares indices are only available from DfT for the PTE areas as a single unit, the English Shires as a whole and then separately for London, Scotland and Wales. Similarly, changes in bus speeds have been estimated by reference to changes in staff productivity identified in our parallel *Bus Industry Performance* report. Some data on speed changes in London since 2014 has been published by TfL and this has been incorporated into the model (see Table 61 above). In the absence of detailed local data, this measure can only act as a proxy, but the figures do give us an indication of overall trends.

The areas discussed are:

- The PTE Areas – combining results from Tyne & Wear, West Yorkshire, South Yorkshire, Greater Manchester, Merseyside and West Midlands.

- English Shire Areas – combining results from the shire areas of government office regions

- Scotland

- Wales

13.2   The PTE Areas

Across the PTE areas, the modelling work suggests that patronage between 2010 and 2019 would fall by 16% from 1,164 million journeys to 891 million. In fact, the number has fallen to 934 million. The variance between modelled and actual is therefore 4.6%. This suggests that other factors such as marketing, quality changes or ticketing improvements, may have mitigated a fall that would otherwise have taken place.

The graph at Figure 13-1 below shows how the mix of factors which account for the change would be made up, given the assumptions made. As can be seen, six factors have, to some extent or another, contributed to a fall in patronage, including:

- the growth in car ownership

- changes in the use of concessionary passes

- real-term fare increases for paying passengers

- reductions in service levels

- slower bus speeds

- changes in shopping habits (as measured by retail footfall).

The modelling work suggests that these falls have been mitigated by economic growth, employment and population changes, by changes in petrol prices (earlier price rises offsetting subsequent sharp falls) and other mitigating factors.

Figure 13-1: Bridging the Gap - Changes in Bus Demand since 2010, PTEs

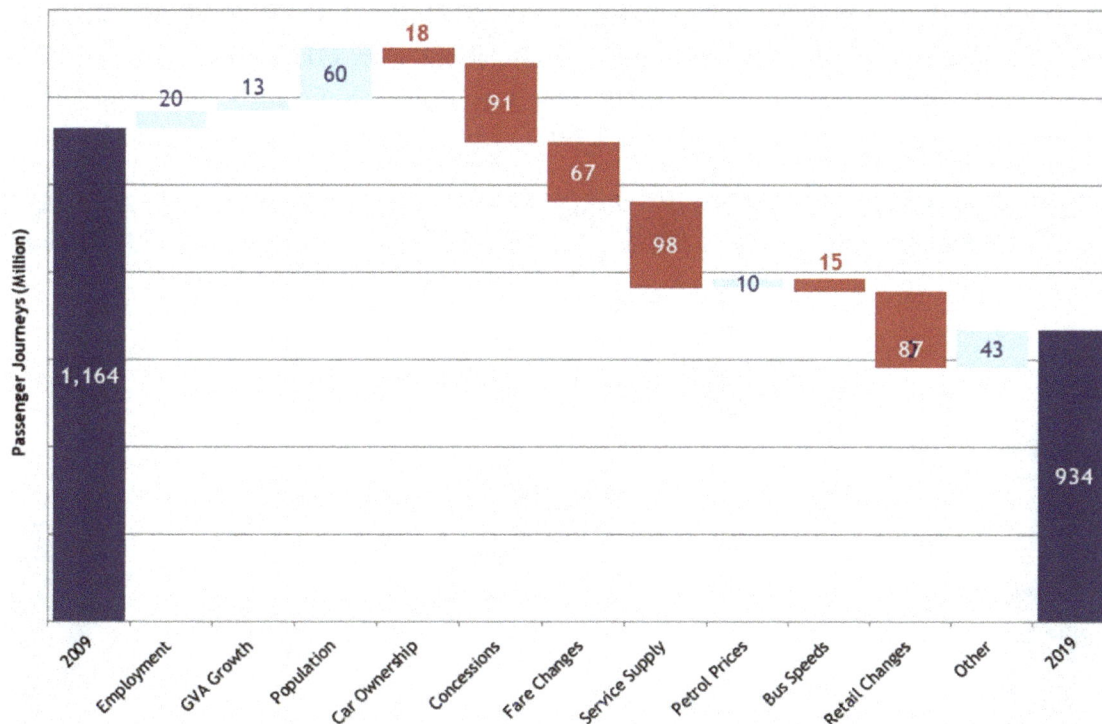

This work suggests that the largest factor contributing to the decline is the reduction in service supply that have taken place, primarily as a result of reduced support budgets from local authorities. Second comes the reduction in concessionary travel and third was the ongoing tendency for bus fares to rise by more than inflation. As we noted earlier, operators are faced with both cost increases and falls in demand for other reasons, and in the absence of operating subsidy from the taxpayer, this gives them no alternative but to increase prices. This is a trap which the industry has faced for more than 50 years under all regulatory regimes and whether publicly or privately owned.

The falls in demand for concessionary travel have been driven by the reduction in the number of eligible people and the increase in the number of older people holding a driving licence and owning a car. Both were discussed in Chapter 5.

13.3   English Shire Areas

In the Shire areas, the modelling shows a 25.8% reduction in patronage over the 2010-2019 period, suggesting that demand would fall to 985 million journeys. The actual result was a figure of 1,213 million, with a variance of 18%. The disparity is believed to be caused by the wide variations in performance in the Shire areas, with 23% of local transport authorities showing growth since 2009/10, and 43% showing growth in 2018/19 (see section 9.5 above). Areas such as Bristol have seen patronage increases of over 50% whilst others such as Warrington and Stoke-on-Trent have seen falls of over 40%. It is clearly impossible for a model operating at this level of aggregation to capture such variations.

The components of the fall are shown in Figure 13-2 below.

In this case, the analysis suggests that the biggest contributor to the decline has been the reductions in concessionary demand, followed by changes to shopping patterns with the decline of the High Street in favour of online shopping. Ongoing increases in car ownership is the third largest reason, the modelling suggests, followed real-term fare increases. The other major driver of decline has been cuts in service supply, again primarily driven by cuts in the provision of supported services by local authorities. These have been offset by changes in population and employment along with GDP growth. Overall, the model suggests that bus speeds have a net benefit too, with some early improvements in productivity as a result of the recession not yet completely negated by subsequent falls.

Figure 13-2: Bridging the Gap - Changes in Bus Demand since 2010, Shires

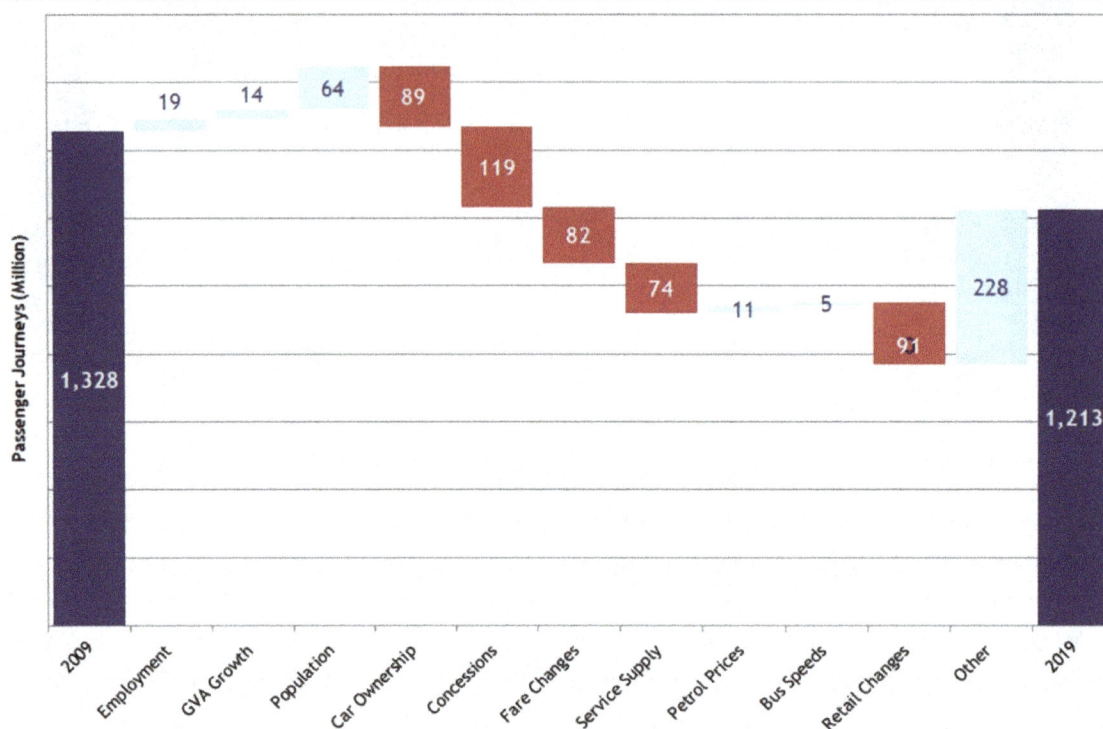

13.4   Scotland

The model suggests a fall of 18.5% in demand in Scotland to a figure of 395 million, whereas in fact the actual fall was larger, down to 380 million. This is a variance of 3.9%.

As noted in section 10.3 above, however, there has been marked variation between different parts of Scotland in recent years. Reviewing the regional figures in Table 48 on page 101 above, it is clear that a substantial chunk of the fall in patronage since 2010 has been concentrated in the South West and Strathclyde area. The North East, Tayside and Central region has also seen some fall, mainly driven by the economic slowdown in the North East after the falls in the oil price. Other parts being either steady or slightly up. One cause for this seems to have been a much higher than average fall in bus speeds in Glasgow[8].

The decline which the model suggests is shown in Figure 13-3 below. Here, the picture is dominated by service cuts, followed by changes to shopping patterns. Falls in demand for free concessionary travel seem to be the third most important factor. As in other parts of the UK, these have been offset to some extent by economic and demographic changes.

---

[8] See The Impact of Congestion on Bus Passengers, *Professor David Begg. Published by Greener Journeys, Autumn 2016*

In Scotland, there have been no changes to the qualification criteria for the over 60s to receive a free pass, but even so passenger demand from concessionary pass holders has fallen.

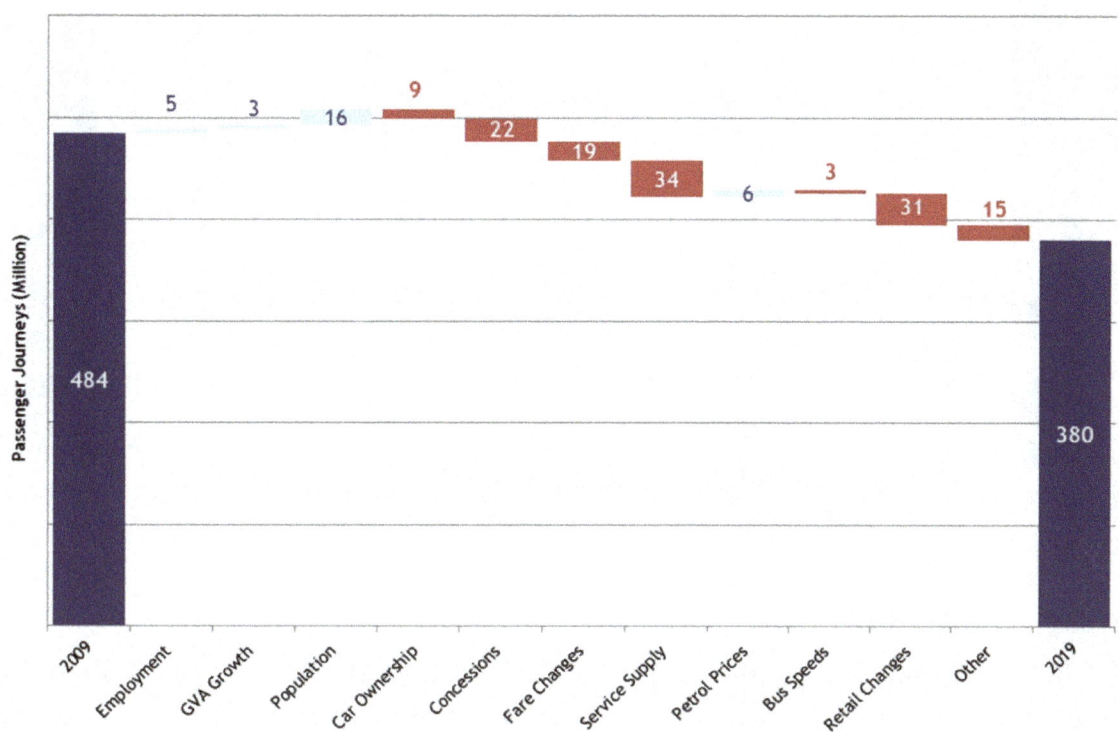

Figure 13-3: Bridging the Gap - Changes in Bus Demand since 2010, Scotland

13.5   Wales

In Wales, the model suggests a fall of around 20.9% between 2010 and 2019, to a total of 99 million journeys. In fact, demand has fallen to 102 million journeys, so that the variance is 3.3%.

The gap seems have arisen primarily amongst concessionary passengers, with the variance on these journeys being 5% compared with paying passengers at 1.9%.

The analysis of the changes is shown in Figure 13-4 below. The analysis suggests that, as with other areas, the largest contributor to the reduction has been reductions in demand for concessionary travel, closely followed by reductions in service supply. Other factors contributing to the mix include changing retail habits, fare changes, car ownership growth, and lower bus speeds. Some growth in employment, population and overall output has offset the downward pressure, as has the overall movement in petrol prices across the period.

Figure 13-4: Bridging the Gap - Changes in Bus Demand since 2010, Wales

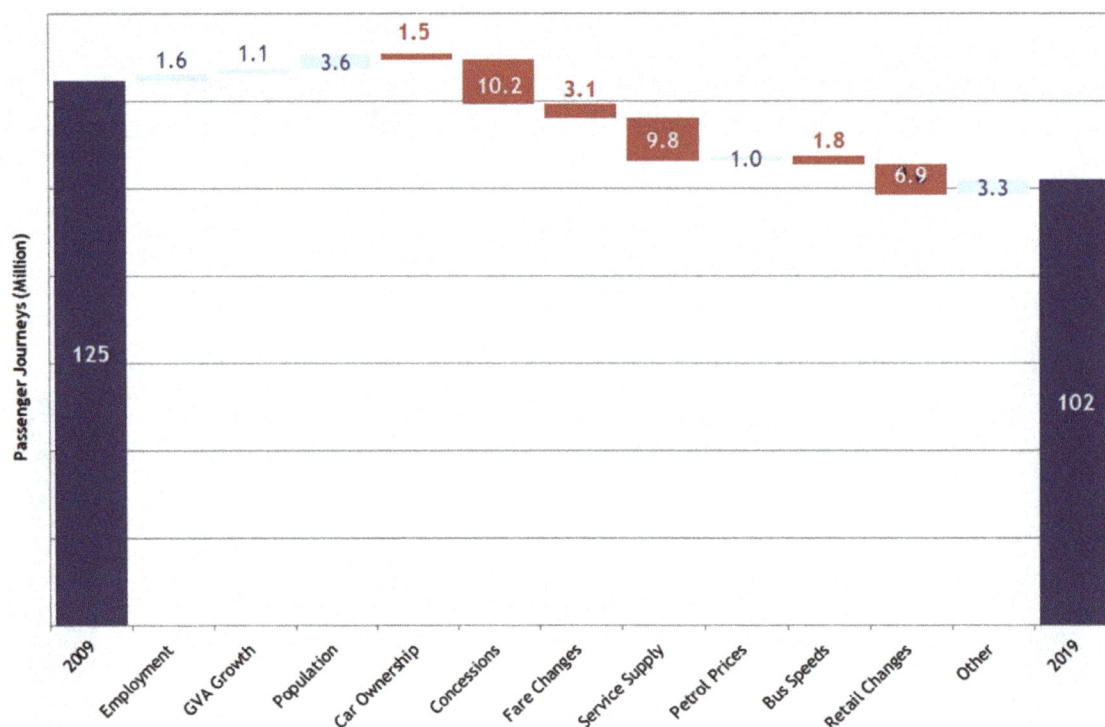

Figure 13-4: Bridging the Gap - Changes in Bus Demand since 2010, Wales

13.6   London

In London, the model suggests increase of 13.6% in demand between 2010 and 2019 to a total of 2,531 million journeys. In fact, there has been a small fall over the period of 1.4%. This is a variance of 15.2% between modelled and actual.

As in Wales, the gap seems to have arisen in the forecast for journeys by fare paying passengers, together with youth and other passes. The numbers for holders of free concessionary passes for elderly and disabled people are only 2.1% apart.

As can be seen, the work suggests that the main drivers of an increase in demand have been demographic and economic, with the recorded fall in car ownership also having an effect. The growth in concessionary demand already discussed is also there, alongside a small net effect from the overall change in petrol prices.

Offsetting these have been reductions prompted by real-term fares increases and cuts in service supply. Changes in shopping habits may also have had an effect, as have changes in bus speeds – though these may be understated. As indicated in the discussion at the end of Chapter 12, the strong performance of the London economy, growth in employment and population would have pointed to continuing increases in demand for travel. This suggests that the growth in congestion and a move towards more home working has driven the decline.

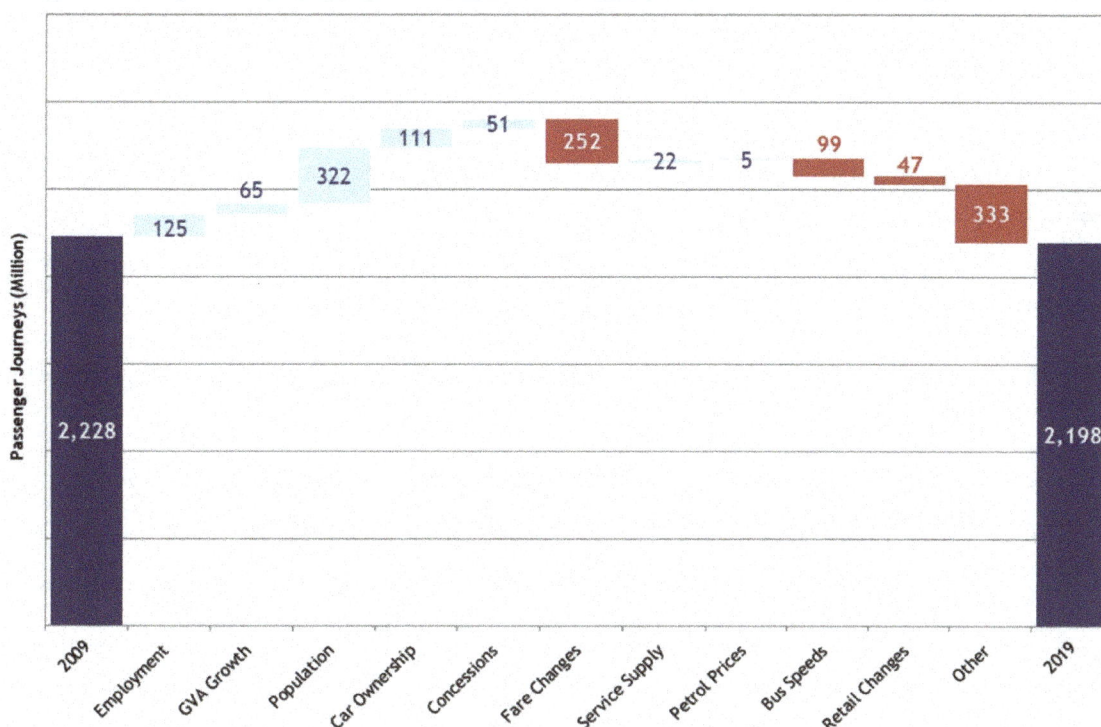

Figure 13-5: Bridging the Gap - Changes in Bus Demand since 2010, London

The chart shows Passenger Journeys (Million) as a waterfall from 2009 to 2019:

- 2009: 2,228
- Employment: 125
- GVA Growth: 65
- Population: 322
- Car Ownership: 111
- Concessions: 51
- Fare Changes: 252
- Service Supply: 22
- Petrol Prices: 5
- Bus Speeds: 99
- Retail Changes: 47
- Other: 333
- 2019: 2,198

13.7 Conclusions

This chapter has illustrated how the various drivers of demand for bus services can influence the numbers of passengers carried in different areas. As with previous work on this subject, the results of a dynamic simulation over a five or six-year period are surprisingly close to what has actually happened on the ground in most cases.

It is particularly noteworthy that the same model using the data from the same sources can predict both the falls in demand that have occurred in some parts of the country and the increases that have been seen in London. This does once again suggest that the different outcomes that have been seen in the capital are much more to do with specific economic and demographic circumstances than with the different regulatory regime.

It is important, though, to emphasise that this sort of work is not purely deterministic. Looking forwards, it is indeed possible to look at forecasts of economic growth, numbers of jobs and numbers of people, and to draw conclusions as to what these *might* mean for the bus industry, other things being equal. Knowledge and understanding of such trends will be of immense help in planning for the future – even in the uncertain times we now face during and after the COVID-19 epidemic.

However, decisions made about service supply, quality of provision and factors which influence bus speeds will all affect the outcome – and it is those which managers need to focus on, either effecting change themselves or influencing those such as Highway Authorities which can deliver it.

Chapter 14: Looking to the Future

14.1   Introduction

It has been suggested in previous editions of *Bus Industry Monitor* that changes in policymaking required to assist with the reversal of the decline in bus patronage were actually happening.

Whilst this might have been true in the years up to 2010, the economic crisis of 2007/08 and its aftermath have changed the situation dramatically in several ways:

- Cuts in spending by local authorities have resulted in major reductions in supported networks

- Cuts to BSOG and concessionary fares reimbursement levels have made bus services less profitable to operate, further reducing service levels at the margin

- The economic problems have focused consumer attention away from environmental issues, so making people less aware and less willing to change behaviour

- Governments have been reluctant to increase fuel duties, partly because of low growth in household incomes and partly because of volatility in the price of oil, thus widening the gap between the cost of public and private transport.

Some of this may be reversed in the coming years, particularly in view of the apparently intractable and growing problems of air quality in urban centres.

As has already been seen, demand for transport, as measured by the Government in passenger kilometres, has more than trebled in the last half-century, totalling 808 billion in 2018. This was the last year for which figures are available at the time of writing. 83% of that demand was met by private car, van and taxi, making a total of 673 billion passenger kilometres. This compared with 35.3 billion for buses and coaches.

The dip caused by the recession passed: after several years of stuttering, demand resumed growth. The previous high of 792.2 billion passenger kilometres, achieved in 2007, was passed in 2016, and the total figure for both 2017 and 2018 exceeded 800 billion for the first time.

At the turn of 2020, it was still expected that there would be a continued long-term growth in car ownership and use, with a consequent rise in the level of road traffic. This had been given expression in the 2018 National Road Traffic Forecast, which envisaged growth of between 17% and 51% between a base year of 2015 and 2050.

Then the COVID-19 virus hit, the lockdown was imposed in the third week of March and everything changed: overnight, demand for transport collapsed.

The impact of the virus and the virtual shutdown of the world economy will undoubtedly have severe consequences on the economy, on government finances and consequently on the demand for transport. In this chapter, we will consider these first, and then discuss the question of climate change, though in the knowledge that COVID-19 may well have massively changed the terms of that debate as well.

14.2   COVID-19 and the Consequences for Bus Demand

In trying to assess the consequences of virus crisis on bus demand, I have considered three headings to consider how things might develop after the initial lockdown period ends. They are:

- Social distancing

- Lifestyle Changes

- The Economy

14.2.1   Social Distancing

I have called the first phase "Social distancing". During this period, restrictions will be progressively lifted and economic life will restart in limited form, but it can be expected that not all venues will reopen, and there will be some unwillingness to leave home – especially amongst older people and those with underlying health conditions. Universities and colleges may not reopen fully until the autumn. Those who are obliged to travel to and from work may prefer to do so in their own car or a taxi rather than sharing space on public transport. Social distancing on public transport will reduce capacity dramatically.

This phase is likely to last several months and will not be fully over until a vaccine and/or drug treatment can be developed and manufactured. Meanwhile, alterations to behaviour and the economic consequences of the crisis will start to play out.

14.2.2   Lifestyle Changes

It seems likely that some of the habits that we have been forced to adopt during the crisis will continue after it is over – especially where these merely accelerated trends that were already under way. These changes are likely to affect all journey purposes:

- Work – more home working and video conferencing, resulting in fewer commuting and business trips

- Shopping – more online shopping resulting in fewer trips and more closures, making high streets less attractive

- Leisure – more home entertainment and streaming, home food deliveries – resulting in fewer trips to theatre, cinema, restaurants and possibly sporting venues.

- Personal business – acceleration of the move of financial services transactions online, coupled with a similar move to remote consultations and diagnostics in the NHS. Both might be expected to reduce demand for personal business and escort trips.

- Education – a move to more online teaching may be accelerated, driven both by demand and the need to cut premises costs.

14.2.3   The Economy

There is little consensus amongst economists as to the nature and extent of the damage that the crisis will have to done to the economy. Some argue that there will be a quick bounce back to activity levels seen before the lockdown; others suggest long-lasting damage. There is a consensus that public spending will be under pressure and that higher taxation is likely to be necessary to service and repay the huge debts that will have been incurred.

Looking again at the five principal purposes for travel by bus, it is possible to identify the threats to each:

- Work and Business – fewer jobs and lower prosperity will impact on the number of commuting and business trips

- Leisure – lower disposable income may be expected to reduce travel demand

- Shopping – reduced footfall as volumes fall and as shop closures make High Streets less attractive places to visit

- Personal – pressure on costs will drive more branches in financial services to close, so reinforcing the trend to drive financial services, legal services and some medical work online.

- Education – spending cuts and reduced demand lead to reducing numbers of students and fewer university places.

14.2.4   Summary

The expectation must be that a combination of some or all these movements would have serious consequences for the current level of demand for bus travel. Given the nature of the business and the marginal nature of much of its existing level of supply, further service cuts or fare increases would be inevitable without continued government intervention.

An illustration of the points discussed in this section can be seen at Figure 14-1 below.

At its core is a doughnut graph of the estimated proportion of bus trips by journey purpose, extracted from the National Travel Survey 2018 but slightly simplified.

Outside the doughnut are three concentric circles, representing the three types of risk discussed above, which we categorised as "social distancing" (inner), "lifestyle changes" (middle) and "the economy" (outer). Within each segment of each journey purpose, the possible cause of a fall in demand as a result of each risk is noted.

The list is not exhaustive, and some will inevitably overlap (for example, declines in leisure spending and high street shopping would inevitably reduce the number of jobs in hospitality and retail, further impacting on demand for travel to work). In current circumstances, any modelling of potential outcomes is very uncertain indeed.

As a rough guide, using the journey purpose numbers, outline calculations on a set of "reasonable" forward assumptions for each journey purpose suggests that a period of social distancing could be expected to run at around 55% to 60% below previous demand levels, whilst a return to "normal" would be at a level between 18 and 26 per cent lower than before the lockdown. Even getting to those levels would depend crucially on the extent and speed of the recovery in the wider economy.

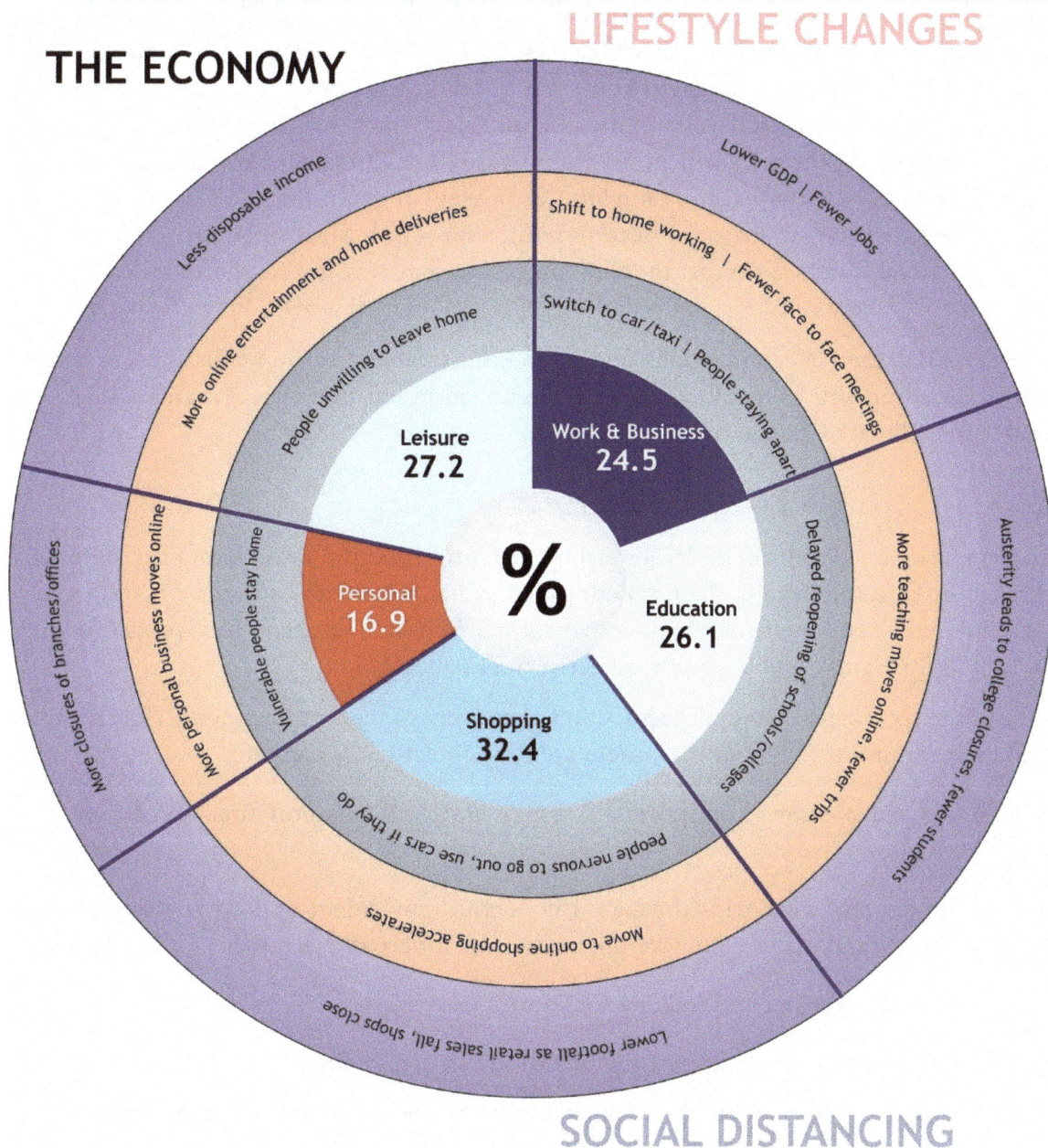

Figure 14-1: Bus Industry Circles of Risk: COVID-19 and its Aftermath

THE ECONOMY

LIFESTYLE CHANGES

Lower GDP | Fewer Jobs

Shift to home working | Fewer face to face meetings

Less disposable income

More online entertainment and home deliveries

Switch to car/taxi | People staying apart

People unwilling to leave home

Leisure
27.2

Work & Business
24.5

%

More closures of branches/offices

More personal business moves online

Vulnerable people stay home

Personal
16.9

Education
26.1

Delayed reopening of schools/colleges

More teaching moves online, fewer trips

Austerity leads to college closures, fewer students

Shopping
32.4

People nervous to go out, use cars if they do

Move to online shopping accelerates

Lower footfall as retail sales fall, shops close

SOCIAL DISTANCING

14.3 Climate Change Policy

14.3.1 The Issues

The prospects of a huge increase in demand for public transport services are growing as increasing emphasis is placed on modal shift as a means of reducing carbon emissions from transport. Both the Committee on Climate Change (CCC) and Friends of the Earth (FoE) have been urging the government for some time to do more to promote modal shift in addition to policies to encourage a shift from internal combustion engines to electric vehicles.

The government's fifth carbon budget, drawn up by the CCC, was given statutory effect in 2016, and covers the years 2028 to 2032. The budget includes a four megaton reduction in CO_2 ($MtCO_2$) to be achieved by shifting 5 per cent of car km taken in trips of the shortest length (<4 miles for bus, <2 miles for cycling and <1 mile for walking) to bus, cycling and walking. In drawing up the budget, the committee further suggested that reducing car distance travelled by 10% could lower emissions by 6 $MtCO_2$ by 2030.

There has been concern amongst CCC members and lobby groups for some time that DfT policies are not fully aligned with such a target, with little or no indication of any measures to deliver modal shift in their approach, which has been focused solely on decarbonisation of the vehicle fleet. In October 2018, CCC Chair Lord Debden wrote to then Transport Secretary Chris Grayling, expressing concern that DfT policies had not been adapted to contribute to the need to reduce car travel. Meanwhile, the Department's National Road Traffic Forecast in 2018 suggested that road traffic increases of between 17% and 51% between 2015 and 2050 - with consequential growth in traffic congestion.

In June 2019, the UK's obligations under the Paris treaty on carbon emissions, to achieve net zero carbon emissions by 2050, were enshrined in law under new regulations passed under the Climate Change Act 2008.

14.3.2   Transport Decarbonisation Plan

In response, the Department for Transport has promised the development of a detailed Transport Decarbonisation Plan, scheduled to be finalised by the autumn of 2020, and to this end published a consultation document, *Decarbonising Transport – Setting the Challenge*, in March 2020, just as the coronavirus crisis was taking hold.

One of the six strands set out in the document concerns "accelerating modal shift to public and active transport". Under that, there are four objectives, which are:

- Help make public transport and active travel the natural first choice for daily activities

- Support fewer car trips through a coherent, convenient and cost-effective public network; and explore how we might use cars differently in future

- Encourage cycling and walking for short journeys

- Explore how to best support the behaviour change required.

Later in the document, the Department restates modal shift as one of its priorities:

"Accelerating modal shift to public and active transport:

We want public transport and active travel to be the natural first choice for our daily activities. An important aspect of reducing emissions from transport will be to use our cars less and be able to rely on a convenient, cost-effective and coherent public transport network. For those able to do so, we would like cycling and walking to be the easy and obvious choice for short journeys. We are already exploring how we can use vehicles differently, such as through shared mobility. New technologies and business models may help facilitate modal shift, such as Mobility as a Service platforms. This will require behavioural changes and we will consider how government and others can support this shift through infrastructure and encouraging those forms of travel."

14.3.3   The Scale of the Challenge

As we have emphasised before – indeed throughout the life of the *Bus Industry Monitor* project – achieving modal shift is not easy and the scale of the change needed tends not to be widely understood. With car demand exceeding bus demand by a factor of more than ten, any switch from car to bus has a major effect on the latter.

As we have seen, total demand for travel in the UK in 2018 was 808 billion passenger km, of which 672 billion (83%) was car-based. The total for bus and coach was 35.3 billion, less than five per cent of the total, with another 80.5 billion by rail. Thus, each one per cent of demand shifted from car to bus would represent a 19.1% increase in bus demand. In other words, achieving even the CCC's modest 5% target could result in up to 95.4% more demand on the buses (or 50% on the trains), whilst a 20% shift would imply increases of 381% and 199% respectively.

Table 63 shows the calculations, which are illustrated in graph form at Figure 14-2 below.

Table 63: Implications of Modal Switch from Car to Bus

Size of Modal Switch from Car (% of existing car demand)	Volume switched to Bus (Billion Passenger Kilometres)	Proportion of Existing Bus Demand (%)
1%	6.73	19.1%
5%	33.64	95.4%
10%	67.27	190.7%
15%	100.91	286.1%
20%	134.54	381.5%

Source: PTIS Analysis of figures from Transport Statistics Great Britain 2019, Department for Transport.

Figure 14-2: Modal Shift from Car - Public Transport Demand Change

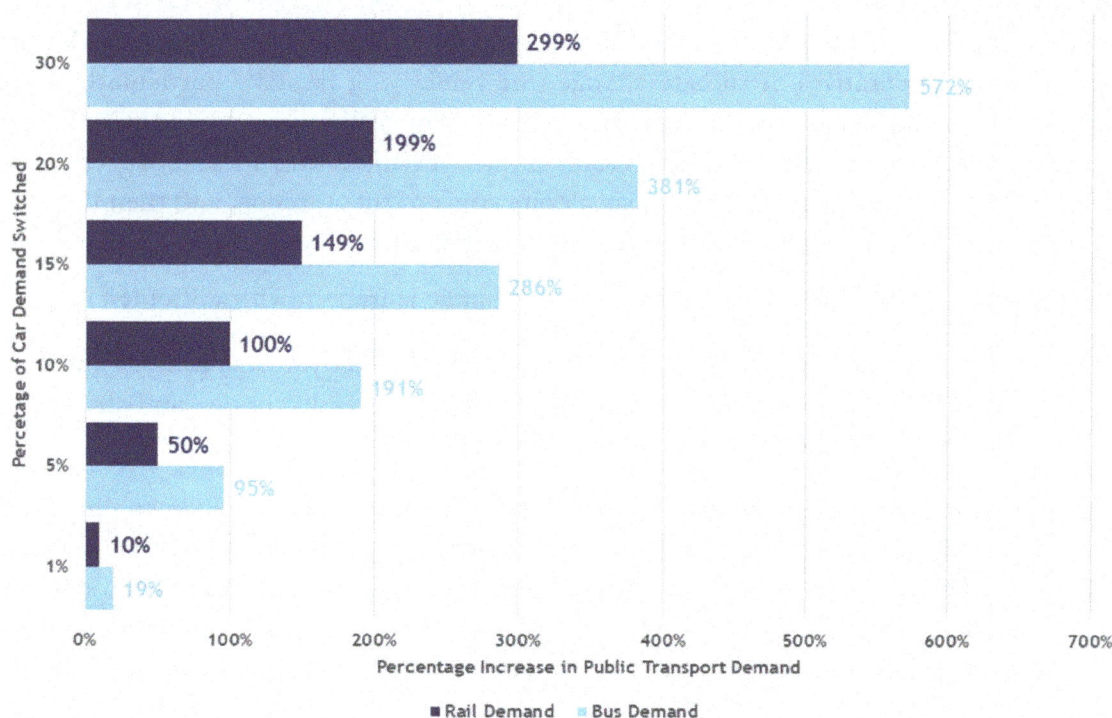

On the buses, the early stages of a major shift could be accommodated by increasing average loads on existing vehicles. Extensive use of priority measures and a reduction in traffic could free bus services from congestion, so boosting the capacity of the existing fleet through better utilisation. Some operators have suggested in the past that this could increase capacity in urban areas by 10 per cent or more.

Once that resource had been exhausted, expansion of services and fleets would be required - with huge increases in the demand for labour. The results would also be immense in terms of the increases that would be needed. This would include new vehicles (or old ones retained for longer), funding requirements and additional staff and management.

Staffing is crucial and will strike a particular chord with managers who have had to deal with past staff shortages in many parts of the country. These have been very difficult, as the resulting inability to provide a reliable service does severe harm to the market position of the bus. Expansion on a large scale would undoubtedly exacerbate recruitment difficulties.

This does not mean that the industry could not cope with expansion – especially bearing in mind the falls in average load discussed earlier, and the significant expansion achieved in London before and after the introduction of congestion charging. However, it would not necessarily be a simple matter, even if the public consent or the political will existed to implement the measures necessary to achieve such a major shift.

14.4   Final Words: Author's Comment

This report set out to provide information on the latest state of the market for the bus product in Great Britain, using latest statistics from a variety of different sources. The second task was to unpick those overall statistics and try to understand the various items that have been driving demand – the jigsaw of influences we identified in Figure 1-1. The third task was to quantify the various pieces of the jigsaw and then reassemble the overall picture. Finally, using that knowledge, we have attempted to understand what the future might look like.

Even before the COVID-19 crisis, it did not look good: ahead of any major modal shift driven by the imperatives of climate change, the underlying trends – particularly higher car ownership and use leading to increased levels of congestion, the move towards more home working and the shift to online retail. As we saw in section 14.2 above, there is a clear likelihood that the lockdown will accelerate some of these trends, and then there will be the downturn that might be driven by any economic slowdown.

Relying on major modal shift driven by climate change is fraught with difficulty:

- We do not know whether and to what extent the (as yet unspecified) measures needed to drive modal shift will be acceptable to the electorate (and the tabloid press) and therefore deliverable politically

- There is a risk – even likelihood – that climate change will slip back down the agenda as it did during and immediately after the financial crisis of 2008, emphasis being placed instead on economic recovery. A fiscal approach to encouraging modal shift, for example through increases in fuel taxation or road user charging, is most unlikely to be politically feasible at a time of low and stagnant household incomes

- Social distancing may well, at least in the short to medium term, drive customers away from public transport and back to their cars

On the other hand, we know that many places have met and overcome the challenges in recent times: as we saw in section 9.5 above, 38 of the 88 (43%) local transport authorities in England saw some measure of bus patronage growth during 2018/19 in the areas, with many having a record of growth across a five-year period, and 23 of them can show growth over the last decade.

This suggests that success can be delivered, and with it a huge range of benefits to the whole community – economic, environmental, commercial and social. Despite the imponderables with which we are currently grappling, this is still a prize worth fighting for.

Lightning Source UK Ltd.
Milton Keynes UK
UKHW050832260522
403559UK00004B/43